Technology Management and International Business

Technology Management and International Business
Internationalization of R&D and Technology

Edited by
Ove Granstrand
Chalmers University of Technology, Sweden

Lars Håkanson
The Stockholm School of Economics, Sweden

Sören Sjölander
Chalmers University of Technology, Sweden

JOHN WILEY & SONS
Chichester · New York · Brisbane · Toronto · Singapore

Copyright © 1992 by John Wiley & Sons Ltd
Baffins Lane, Chichester
West Sussex PO19 1UD, England

All rights reserved.

No part of this book may be reproduced by any means,
or transmitted, or translated into a machine language
without the written permission of the publisher.

Other Wiley Editorial Offices

John Wiley & Sons, Inc., 605 Third Avenue,
New York, NY 10158-0012, USA

Jacaranda Wiley Ltd, G.P.O. Box 859, Brisbane,
Queensland 4001, Australia

John Wiley & Sons (Canada) Ltd, 22 Worcester Road,
Rexdale, Ontario M9W 1L1, Canada

John Wiley & Sons (SEA) Pte Ltd, 37 Jalan Pemimpin #05-04,
Block B, Union Industrial Building, Singapore 2057

Library of Congress Cataloging-in-Publication Data

Technology management and international business :
 Internationalization of R&D and technology / edited by Ove
Granstrand, Lars Håkanson, Sören Sjölander.
   p. cm.
 Includes bibliographical references and index.
 ISBN 0-471-93425-9
 1. Technology—Management. 2. Technology—International
cooperation. I. Granstrand, Ove. II. Håkanson, Lars.
III. Sjölander, Sören.
T49.5.T445 1992   91–45187
658.5'7—dc20   CIP

British Library Cataloguing in Publication Data

A catalogue record for this book is
available from the British Library.

ISBN 0-471-93425-9

Typeset in 10/12pt Palatino by Text Processing Dept, John Wiley & Sons Ltd, Chichester
Printed and bound in Great Britain by Biddles Ltd, Guildford and King's Lynn

Contents

Contributors ix

Preface xi

Abbreviations xiii

1 Introduction and Overview 1
Ove Granstrand, Lars Håkanson and Sören Sjölander
1.1 Introduction 1
1.2 Background and perspectives 1
1.3 International R&D in multinational corporations 5
1.4 Present and future of the research agenda 14
1.5 References 15

2 Multinational Enterprises and the Globalization of 19
Innovatory Capacity *John H. Dunning*
2.1 Introduction 19
2.2 Innovatory capacity by country: The location of capacity 20
2.3 Innovatory capacity by firms: The ownership of capacity 25
2.4 Why do MNEs engage in foreign innovatory activities? 29
2.5 Some locational patterns 32
2.6 The impact of MNEs on innovatory capacity of host countries 34
2.7 Two case studies 39
2.8 The role of government policy 43
2.9 Conclusions 47
2.10 A note on the future 49
2.11 References 49

3	**Large Firms in the Production of the World's Technology: an Important Case of Non-Globalisation** *Pari Patel and Keith Pavitt*	53
3.1	Introduction	53
3.2	The data set, its advantages and limitations	55
3.3	Large firms in the production of the world's technology	59
3.4	Large firms in home countries' technological activities	62
3.5	Large firm performance and country performance	64
3.6	Conclusions	71
3.7	References	72

4	**The Internationalisation of Technological Activity and its Implications for Competitiveness** *John Cantwell*	75
4.1	Introduction	76
4.2	The data: how patent statistics can be used	78
4.3	The internationalisation of technological activity amongst the world's largest industrial firms; its geographical and sectoral composition	79
4.4	The international location of innovative activity and the technological specialisation of host countries	83
4.5	The interaction between the internationalisation of technological activity and national competitiveness: the consequences for British firms and industries	89
4.6	Changes in the technological specialisation of MNCs	92
4.7	Conclusion	94
4.8	References	95

5	**Locational Determinants of Foreign R&D in Swedish Multinationals** *Lars Håkanson*	97
5.1	Introduction	97
5.2	The database	99
5.3	A typology of foreign R&D units	100
5.4	Locational determinants	105
5.5	Empirical tests	110
5.6	Summary and conclusions	113
5.7	References	114

Contents — vii

6	**Business Culture and International Technology: Research Managers' Perceptions of Recent Changes in Corporate R&D** *Mark Casson, Robert D. Pearce and Satwinder Singh*	**117**
6.1	Introduction	118
6.2	The changing scope of corporate R&D	118
6.3	Trends in organisational structure	123
6.4	The practice of R&D management—evidence from interviews	125
6.5	Government policy	129
6.6	Government funding	131
6.7	Appendix	133
6.8	References	135

7	**Internationalisation of Research and Development among the World's Leading Enterprises: Survey Analysis of Organisation and Motivation** *Robert D. Pearce and Satwinder Singh*	**137**
7.1	Introduction	138
7.2	The survey	142
7.3	Results	143
7.4	Conclusions	160
7.5	References	162

8	**Management of International R&D Operations** *Arnoud De Meyer*	**163**
8.1	Introduction	163
8.2	The research basis	164
8.3	Why internationalize R&D?—the traditional reasons	166
8.4	Validation of these models	168
8.5	Technical learning as a factor of internationalising R&D	169
8.6	The planning and control contribution to technical learning	170
8.7	Networking as a core element of the organisation	173
8.8	Communication	176
8.9	Conclusion	178
8.10	References	178

9 Internationalization and Diversification of Multi-technology Corporations 181
Ove Granstrand and Sören Sjölander

9.1	Introduction	182
9.2	Technology management perceptions in Japan, Sweden and the USA	184
9.3	Internationalization of technology acquisition	187
9.4	Diversification	190
9.5	Internationalization and diversification: The case of cellular systems and terminals	192
9.6	Internationalization, diversification and technology—a tentative model	197
9.7	Summary and conclusions	201
9.8	References	204
9.9	Appendix: Assertions about technological progress, technology diversification, internationalization and competition	206

10 International Collaborative Ventures and US Firms' Technology Strategies *David C. Mowery* 209

10.1	Introduction	209
10.2	International collaborative ventures: Definition and growth	211
10.3	Comparing joint ventures and other channels for the exploitation of technological and other firm-specific assets	214
10.4	Causes of increased reliance on joint ventures	218
10.5	Influences on the structure of international collaborative ventures	223
10.6	Managing international collaborative ventures	224
10.7	Conclusion	229
10.8	References	230

11 Summary and Implications 233
Ove Granstrand, Lars Håkanson and Sören Sjölander

11.1	Summary and some general trends	233
11.2	Managerial implications	239
11.3	Policy implications	242
11.4	A speculative outlook on the future	247

Index 251

Contributors

JOHN CANTWELL   University of Reading, Department of Economics, Whiteknights, PO Box 218, Reading, Berkshire RG6 2AA, England, tel: +44 - 734 - 87 51 23, fax: +44 - 734 - 75 02 36

MARK CASSON   University of Reading, Department of Economics, Whiteknights, PO Box 218, Reading, Berkshire RG6 2AA, England, tel: +44 - 734 - 31 82 27, fax: +44 - 734 - 75 02 36

ARNOUD DE MEYER   European Institute of Business Administration, INSEAD, Boulevard de Constance, 77305 Fontainebleau Cedex, France, tel: +33 - 1 - 60 72 40 00, fax: +33 - 1 - 60 72 42 42

JOHN H. DUNNING   University of Reading, Department of Economics, Whiteknights, PO Box 218, Reading, Berkshire RG6 2AA, England, tel: +44 - 734 - 31 81 59, fax: +44 - 734 - 31 03 49

OVE GRANSTRAND   Chalmers University of Technology, Department of Industrial Management and Economics, 412 96 Göteborg, Sweden, tel: +46 - 31- 772 12 09/12 02, fax: +46 - 31- 772 12 37

LARS HÅKANSON   Institute of International Business, Stockholm School of Economics, PO Box 6501, 113 83 Stockholm, Sweden, tel: +46 - 8 - 73 690 00, fax: +46 - 8 - 32 65 24

DAVID C. MOWERY   University of California at Berkeley, Walter A Haas School of Business, 350 Barrows Hall, Berkeley, CA 94720, USA, tel: +1 - 510 - 64 39 992, fax: +1 - 510 - 64 22 826

PARI PATEL   Science Policy Research Unit (SPRU), University of Sussex, Mantell Building, Falmer, Brighton, East Sussex BN1 9RF, England, tel: +44 - 273 - 68 67 58, fax: +44 - 273 - 68 58 65

Contributors

KEITH PAVITT Science Policy Research Unit (SPRU), University of Sussex, Mantell Building, Falmer, Brighton, East Sussex BN1 9RF, England, tel: +44 - 273 - 68 67 58, fax: +44 - 273 - 68 58 65

ROBERT D. PEARCE University of Reading, Department of Economics, Whiteknights, PO Box 218, Reading, Berkshire RG6 2AA, England, tel: +44 - 734 - 31 82 27, fax: +44 - 734 - 75 02 36

SATWINDER SINGH University of Reading, Department of Economics, Whiteknights, PO Box 218, Reading, Berkshire RG6 2AA, England, tel: +44 - 734 - 31 82 27, fax: +44 - 734 - 75 02 36

SÖREN SJÖLANDER Chalmers University of Technology, Department of Industrial Management and Economics, 412 96 Göteborg, Sweden, tel: +46 - 31 - 772 12 15/12 02, fax: +46 - 31 - 772 12 37

Preface

This book is an edited selection of revised papers, which were originally prepared for an international research conference titled 'Technology Management and International Business' held in Stockholm 17–20 June, 1990.

A number of themes related to technology management and international business were treated at the conference, but internationalization of industrial R&D and technology was the theme with the largest and most coherent subset of conference papers. These papers presented research results from a number of countries, mainly by European researchers. Although the conference was international, this European bias was perhaps not surprising since a number of European corporations, especially from some of the smaller countries, have historically been the most internationalized in the world, both regarding sales, production and R&D.

It is clear that internationalization of industrial R&D with a few exceptions is a fairly recent phenomenon, which still has not progressed very far in absolute terms. However, there are also some clear and general long-term trends towards an increasing extent and importance of R&D conducted on an international basis. The European corporations and countries with highly internationalized R&D compared to others would then become precursors with possible early mover advantages and disadvantages. At the same time most of the research and writings on foreign R&D (which is then typically called 'overseas R&D') have been of US origin.

Against this background it was felt that a book focusing on internationalization of R&D, drawing on research mainly but not exclusively from a European perspective, would be appropriate. Undoubtedly, this focus meant that several other highly interesting themes and contributions at the conference had to be sacrificed, for example internationalization of production systems and the impact of automation and information technologies in that connection.

The conference was moreover designed to bring together perspectives from economists, engineers and business management professionals as well as questions at the macro and micro level. The experience from this design was positive, although a major issue of divergence among participants concerned the pros and cons of various methods—the appropriateness of statistical data of various kinds versus case studies, surveys versus in-depth interviews, etc. It is easy to side-step such divergence of opinion by arguing in favor of methodological pluralism at an aggregate level, but then—how could a suitable integration of research results be accomplished in general and how could they be translated into useful guidelines for management and policy-makers in specific cases? This is certainly not an issue that only concerns research on the internationalization of R&D. However, in view of the youth of this phenomenon, one cannot, for the time being, expect that much can be said on firm grounds with general validity, except that more recognition of various aspects of the phenomenon is needed and also that more research is needed. As is so often the case in management studies, management research and policy research lag behind best management practice. This applies as well to technology management in general and internationalization of R&D in particular. The conference and this book have been an attempt to catch up a little bit. However, more efforts are needed and, hopefully, this book can also stimulate further work in the area.

Finally, we as editors and conference organizers want to mention and sincerely thank many persons and institutions that have helped us in this effort. The conference was organized jointly by the Department of Industrial Management and Economics at Chalmers University of Technology in Göteborg, Sweden, and the Institute of International Business (IIB) at the Stockholm School of Economics under the auspices of the Institute for Management of Innovation and Technology (IMIT). In this connection we want to thank the present head of IMIT, Professor Bengt Stymne, Anna Karlstedt, IMIT, and Ulla Anson, Chalmers, as well as Robert Nobel and Dr Udo Zander at IIB for their assistance. We also want to thank all the participants at the conference for valuable discussions and contributions. Special thanks go to the sponsors of the conference—Prince Bertil's Foundation at the Stockholm School of Economics, the former National Swedish Board for Technical Development, and finally to Chalmers and IIB. In connection with producing this book, we would first of all like to thank all the contributors. Our thanks also go to Ulla Anson for her general secretarial assistance, to Gullvi Nilsson for her invaluable editorial assistance and language check and to Solwy Andreasson for the layout of this book.

OVE GRANSTRAND
LARS HÅKANSON
SÖREN SJÖLANDER
Göteborg and Linz
October 1991

Abbreviations

A&M	Acquisitions and Mergers
CAD	Computer Aided Design
CAM	Computer Aided Manufacturing
CMOS	Complementary Metal Oxide Semiconductor
FDI	Foreign Direct Investment
FMS	Flexible Manufacturing Systems
HDTV	High Definition Television
IIL	Internationally Interdependent Laboratories
LSI	Large Scale Integrated circuits
LIL	Locally Integrated Laboratory
MNC	Multinational Corporation (Company)
MNE	Multinational Enterprise
MNU	Multinational University
MPC	Multi-Product Company
MTC	Multi-Technology Corporation (Company)
NIC	Newly Industrialized Country
R&D	Research and Development
S&T	Science and Technology
WPM	World Product Mandate
VLSI	Very Large Scale Integrated circuits

Chapter 1

Introduction and Overview

OVE GRANSTRAND, LARS HÅKANSON
AND SÖREN SJÖLANDER

1.1 INTRODUCTION

The present volume represents an effort to define the current state of the art concerning the management and economic impact of geographically decentralized research and development (R&D) in multinational companies (MNCs). This line of research has recently begun to attract renewed interest, in reflection of not only critical new trends regarding the nature and conditions of technological change but also significant developments in the realm of economic and managerial theory.

The aim of this introductory chapter is, first, to outline the empirical and theoretical background and perspective of this new research agenda. Second, it gives a summary and overview of prior research in the area, thus defining some of the starting points for the research presented in later chapters. The concluding section outlines the content and structure of the book. Rather than presenting a pre-digested synthesis of the papers, it attempts to identify some of their implications in terms of critical areas for future research.

1.2 BACKGROUND AND PERSPECTIVES

Concurrent with the accelerating pace of technological change, research and development costs have been rapidly escalating, as have the technical and

---

Technology Management and International Business: Internationalization of R&D and Technology.
Edited by O. Granstrand, L. Håkanson and S. Sjölander. © 1992 John Wiley & Sons Ltd

commercial risks associated with investments in R&D. Moreover, especially in our most dynamic industries, technical development has become increasingly 'science-based' [17]. In the past, industrial R&D could be characterized as almost exclusively 'D', being based on well known scientific and technical principles. Today, an increasing amount of industrial R&D takes place on the frontiers of the scientific disciplines and includes both basic and applied research. In addition, many industries are today characterized by the phenomenon, where successful R&D increasingly requires the simultaneous application of scientific and technical knowledge from hitherto distinct and separate technologies.

Technological changes have profoundly affected also the rules of the game in the international market place. New and more efficient means of transport and communication have created new modes of organization and competition. Through their impact on socio-economic structures and income patterns, they have affected the nature of demand, leading to increasing homogenization of certain international markets, whilst generating wider variety and fragmentation in others. Concurrently, new production technologies have been created that permit smaller production runs, while simultaneously making more efficient use of capital, energy, materials and labor.

In short, technological change is constantly creating and reshaping the opportunities and problems confronting society. In a complex interplay, parallel institutional developments,—e.g. in the areas of trade policy, environmental legislation, and infrastructural investments in education, housing, communication and many other areas—sometimes facilitate, sometimes discourage, the exploitation of the new technological possibilities [15].

Concurrent with these tendencies, recent developments in the microeconomic theory of technological change [e.g. 10, 11, 17, 18, 42] have laid the foundation for an improved empirical and theoretical understanding of the nature of technical change and its impact on society. A new set of concepts and theories has recently become available that help us address the complex issues confronting a society in which technology is the chief driving force of social, economic and cultural change. Economists are gradually beginning to unlock the 'black box' of technology.

The articles collected in this volume were written against the background of these developments. They reflect a common interest in one of the principal agents of technological change: the multinational corporation. MNCs control a vast proportion of the world's scientific and technical resources. Notwithstanding the ongoing and unresolved debate concerning the efficiency with which large multinationals employ these resources, there can be no doubt that their decisions regarding the size, content and direction of R&D programs, as well as the pace, mode and geography of the exploitation of R&D results, vitally affect the international economy.

The aim of this book is to contribute to a better theoretical and empirical understanding of the management of technology in international business.

Introduction and Overview ——————————————————— 3

The driving force behind the book is the belief that such an understanding is required if we are to effectively address the normative issues associated both with the management of innovation and technology in organizationally and geographically decentralized multinational firms, and with the welfare implications of MNC behavior for host and home countries. Mirroring this dual task, the book relates to two distinct—though related—research agendas, both of which have attracted much recent attention and which belong to the forefront of their respective fields. These are briefly outlined in the following two sections.

1.2.1 Management and Organization of Modern Multinationals

The study of international R&D management is strongly related to current theoretical and empirical research on the strategy, organization and management of multinational companies. Here, recent contributions (e.g. [3, 4, 5, 22]) have abandoned the traditional view of multinational management as the task of centrally controlling and coordinating a set of peripheral, and largely independent, national subsidiaries. Abandoning the study of headquarter–subsidiary relationships as an essentially dyadic link, recent theory emphasizes the network character of MNCs and the problems and opportunities encountered in exploiting an international organization as an integrated whole.

In the words of one prominent scholar:

> ...management breaks away from the restricted view that activities for which global scale or specialized knowledge is important must be centralized. They ensure that viable national units achieve global scale by making them the company's world source. Similarly, if important technological advances, or market developments are occurring in locations far from the company's headquarters, they work to secure the cooperation and involvement of the relevant national units in the development of the company's technology, its new products, and even its marketing strategy.
>
> In the integrated network configuration, national units are no longer viewed only as the end of a delivery pipeline for company products, or as implementers of centrally defined strategies, or even as local adapters and modifiers of corporate approaches. Rather, the assumption behind this configuration is that management should consider each of the worldwide units as sources of ideas, skills, capabilities, and knowledge that can be harnessed for the benefit of the total organization.[3, p. 381]

This change of perspective has opened up an entirely new research agenda, with the aim of creating a conceptual and empirical basis, by which to reach normative conclusions regarding both the structural properties of large and diversified multinational organizations and the distinctive management processes and systems required to coordinate and control complex international networks of diverse and differentiated units. Several of the contributions to this volume (Pearce and Singh, Cantwell, De Meyer) make explicit reference to and use of this framework.

Although the tendencies outlined above affect—in different ways—all functional areas, R&D management provides a particularly useful vantage point when exploring the theoretical and practical implications of these changes. First, because of its impact on long run viability and competitiveness, R&D is always a *strategic* function. Second, the creation, exploitation and dissemination of new technology in an international organization require the simultaneous achievement of *efficiency, local responsiveness*, and *worldwide learning and know-how transfer*, i.e. the three management tasks identified by Bartlett and Goshal [4] to be the critical ones in the management of modern transnational corporations. Thus, the study of international management of technology provides significant insights not only regarding appropriate mechanisms and procedures for the R&D function [33]. It also throws light on the wider issues concerning the structures, systems and strategies of modern-day multinationals.

1.2.2 International Trade and the Competitiveness of Nations

Through their emphasis on the geographical location of research and development activities and on the patterns of diffusion of technical knowledge, the articles in this book also relate to the macro-economic effects of MNC behavior. As emphasized by Porter [45], the traditional explanations for the relative competitiveness of individual nations—based on various types of factor advantage due to variations in exchange and interest rates, resource endowments and factor costs, capital availability and labor productivity, management practices and government policies, etc.—are insufficient for explaining international trade patterns and differences in economic growth rates between nations. Porter's own analysis emphasizes the role of companies' home nations as the central sources of the skills and technology needed to sustain competitiveness. Although the international spread of the world's largest multinationals might appear to have made their home base less important, Porter's premise is strongly supported by the analysis of Patel and Pavitt in this volume.

New developments in international trade theories have not only more adequately incorporated factors related to R&D, technology and innovation but also in so doing provided arguments for certain types of protectionism and strategic behavior in certain types of situations [12, 23, 38, 50, 56]. These arguments are especially nurtured by the recognition of the phenomenon of increasing returns, so often associated with R&D and technology. The nature and implications of increasing returns have recently been most prominently developed in the works by Arthur, see [2]. However, the specifics of internationalization of R&D and technology have not yet featured strongly in the works on international trade and competitiveness. New developments regarding international trade and investments in technology and their implications for international competition have to be more emphasized in the years to come.

1.3 INTERNATIONAL R&D IN MULTINATIONAL CORPORATIONS

The role of multinational enterprises in the creation, exploitation and diffusion of new technology has long been recognized and has given rise to a number of important research traditions. These include economic analyses of technology transfer by MNCs to third world countries [40] and the role of technology in international corporate strategy [16]. Furthermore, ever since Hymer, economic theory on foreign direct investment has stressed the role of firm-specific technology as an element in the international growth of enterprises [14, 29]; this is also true of more recent formulations that emphasize the role of the 'internalization' motive [7, 24].

However, the specific focus of this volume—the internationalization of research and development in multinational corporations—has so far attracted only peripheral and passing attention in the literature. In spite of some pioneering contributions (e.g. [8, 9, 46, 47 and to some extent 51]), the topic has remained at the fringes of multinational management research. Although studies of the impact on the host country of foreign direct investment traditionally include analyses of R&D performance in foreign subsidiaries, they have tended not to differentiate between the many various types of tasks subsumed under this title.

The lack of interest regarding the location of R&D in MNCs was undoubtedly a reflection of conventional wisdom which has long held that such activities tended to be highly centralized to the home country, i.e. to the vicinity of the corporate head office and major production units. This assumption was strongly supported by empirical observations, largely based on US data.

However, over the last decade, both casual observation and new evidence have forced a reconsideration of this inherited view [34, 54]. As indicated by US patent data (Cantwell, Patel and Pavitt), large Belgian and Dutch firms performed more of their technological activity outside the home country than inside it; in British, Canadian, Swedish and Swiss firms, foreign shares ranged between 30 and 42%. Although these shares are less than the equivalent share of production in these firms, they do indicate that many MNCs perform considerable amounts of R&D abroad.

Also in the case of US and Japanese multinationals, where R&D has traditionally been much more centralized, overseas technical activities are taking on a new significance. In a recent survey [21], executives in both countries ranked the 'internationalization of R&D' as one of their top priorities—

findings echoed in the results presented by Pearce and Singh.

Traditionally, the geographical location of R&D in multinational companies has been analyzed in terms of a trade-off between the competing 'centripetal' forces for (geographical in contrast to organizational) centralization and the 'centrifugal' forces for decentralization. For the benefit of readers not familiar with the literature in this area, the following sections attempt to synthesize the findings of previous studies.

1.3.1 Forces for Centralization

Several supporting causes for centralization have been advanced in the literature. Some of these provide the reasons why R&D should be concentrated to one or to a limited number of units, regardless of location; others also provide a rationale for home country location.

Some writers stress companies' *need to protect firm-specific technology* [48] as an explanation for centralization of R&D near to the head office or major production units in the home country. According to this line of argument, centralizing R&D to the home country minimizes the risks of unwanted 'leakage' to competitors. Indeed, this argument is sometimes used by corporate executives when considering the establishment of foreign R&D units. However, its validity—which seems to rest on the assumption that home country personnel is generally more trustworthy than host country engineers—remains to be established and appears somewhat dubious.

Nevertheless, it seems obvious that the risk of unwanted diffusion of R&D results increases with the number of persons and organizational units involved. This would suggest concentration of vital R&D activities to as few units as possible, but gives no obvious clue as to their location.

Of greater significance for home country location of R&D is probably the fact that firm-specific technological advantages tend to evolve from, and mirror, *home market conditions* [6, 52]. In order to retain and strengthen such advantages, continued close contact with the domestic market and its customers is likely to be vital. Such conditions may be particularly significant for US multinationals, whose competitive advantages are based on products designed to meet the demand on the vast and technically advanced domestic market. But similar conditions may also apply to companies from smaller European markets. In several Swedish multinationals, close cooperation with technically advanced and technically demanding domestic customers has laid the foundation for the technical innovations on which they have subsequently built their international expansion.

Other types of arguments favoring concentration of resources (without indications as to the optimal location) emphasize the significance of *scale economies in R&D* and the difficulties of reaching 'critical mass' in decentralized laboratories. Such scale advantages have several sources. Necessary but expensive equipment and the need to utilize specific types of scientific

Introduction and Overview

expertise can create indivisibilities and require a certain minimum volume of R&D in order to be economically viable. However, empirical tests as to the presence and influence of scale economies in R&D in retarding international decentralization [25, 27, 30, 31, 55] are inconclusive, partly reflecting the difficulty of obtaining accurate measurements.

Clearly, many types of R&D require a relatively large volume in order to be efficient; other types can be undertaken by one or a few engineers. Scale economies in R&D may certainly motivate the concentration of certain activities to central laboratories, but this need not preclude that others can be decentralized to smaller units. In most industries, R&D focusing on minor adaptations of products and processes can be carried out in small laboratories without loss of efficiency.

Moreover, with modern communication technologies, efficient systems for coordination and control may alleviate—perhaps even nullify—the drawbacks of performing R&D in an internationally decentralized network. Digital Equipment has long used electronic networks in product development permitting the use of resources more or less worldwide and around the clock for various product development projects. Such a network system may give smaller units access to central equipment and expertise, facilitating their speedy and cost-efficient utilization.

A further reason for concentrating R&D to a limited number of units is the wish to *minimize the costs of coordination and control*. Coordination and information exchange are required both between R&D units—to avoid duplication of effort and excessive product differentiation, and to promote cross-fertilization and learning—and between R&D units and other corporate functions, such as production, marketing and top management [53]. Such information exchange is usually of an unstructured nature and tends to involve negotiation, persuasion and common problem-solving, i.e. activities that typically require face-to-face contact. Locating company officials within a limited geographical area reduces the (sometimes quite considerable) costs associated with geographically dispersed activities located in different cultural regions.

The most important reasons for multinationals locating the bulk of their R&D to the home country are 'historical'. When R&D departments were first established, it was natural to locate them on existing company premises, i.e. close to the head office and central factories. As was typically the case with the location of head office functions, explicit locational considerations were rarely made [1].

Clearly, the observed pattern of increased internationalization of R&D indicates that the strong forces for centralization of R&D are often counterbalanced by even greater 'centrifugal' forces. The reasons for performing industrial R&D on an international scale are numerous and varied but can conveniently be classified into two groups: 'Demand-oriented' factors denote circumstances and deliberations inducing companies to locate R&D

abroad in order better to serve foreign national markets; 'supply oriented' factors refer to characteristics in the local foreign environment that enhance the efficiency of R&D by providing e.g. favorable access to skilled technical expertise perhaps at lower cost than available elsewhere, access to foreign universities and other research establishments, etc.

1.3.2 'Demand-oriented' Forces for Decentralization

The establishment of foreign manufacturing subsidiaries typically requires that the company—first through exports, later through a sales subsidiary—has attained a certain position in the local market [27]. Such greenfield establishments tend initially to be totally dependent on parent company technology and technical support. In some subsidiaries, especially in those established on smaller markets with slow growth and slow technical change, such technical dependence can remain for a very long time, perhaps indefinitely.

However, transfer of manufacturing technology from the parent usually requires at least some, sometimes substantial, adaptations of production processes (shorter/longer production runs, other raw material qualities or components) and products to match local demand. When such transfers occur only sporadically, these adaptations can usually be handled by technicians from the parent R&D laboratory. However, when this type of work becomes more frequent, the establishment of a local R&D unit is often advantageous. This is particularly the case when such units can also be entrusted with other technical tasks (customization, technical advice and service), when home country laboratories find it difficult to set aside resources to facilitate technology transfers, or when technical support activities are impeded by geographical distance.

Such *technical support laboratories* are therefore to be found primarily in large manufacturing subsidiaries located in significant markets. They tend to be small and report to the local subsidiary management.

However, technical support activities tend over time to evolve into proper development projects [19, Chapter 6]. The impetus for such development continually appears in the commercial relationships with the local market; specific customer problems that cannot immediately be solved, unexpected results occurring in other types of technical activities that can be commercially exploited, etc. Foreign subsidiary management—at least in profitable subsidiaries—often have strong inducements to perform such R&D, both to strengthen the competitive position and to facilitate recruitment of qualified engineers. Also engineers in foreign subsidiaries foster local R&D.

> Once some technical capability has been established, domestic scientists and engineers are certain to see additional opportunities for improvement. There is an almost irresistible creepage from production engineering upstream into design and development. [51, p. 212]

Sometimes, *government regulations* reinforce the inducements to set up local adaptive R&D. This is the case in the pharmaceutical industry, where clinical tests often need to be carried out in the host market. Similarly, in certain industries, where the government or government-controlled bodies act as major customers—e.g. telecommunications and military equipment—local production and local R&D may be an important argument in negotiations and are frequently a prerequisite for obtaining an order.

The dominating economic rationale for undertaking R&D in these types of subsidiaries is market proximity, i.e. the possibility of performing R&D in cooperation with local customers and of maintaining contact with local market trends and idiosyncratic aspects of demand.

The existence and role of such market-oriented R&D units are relatively straightforward and uncontroversial. Since their main mission is to adapt processes and products to local circumstances, they can be assumed to have a largely positive impact on host economies. Moreover, efficient adaptation to foreign markets can only with great difficulty be carried out in the parent company laboratories. Hence, from a home country perspective, there is no real danger that the work of these units will replace that which otherwise would have been carried out in the home country. Their managerial impact is likely to be limited; since the volume of local R&D is usually small and focused on the local market, the need for coordination, control and information exchange is limited.

1.3.3 'Supply-oriented' Forces for Decentralization

As outlined above, the basic aim of the traditional forms of foreign R&D was to enhance the value of existing parent company technology. Today, a new pattern can increasingly be discerned. As a consequence of the rapidly escalating pace and costs of technological development, and the increasing number of sources of front-line technologies and their concurrent combinations in different products and processes, creating and maintaining technological competitive advantage increasingly require access to a wider range of scientific and technological skills and knowledge than is available in the home market.

For these reasons, the missions of foreign R&D units have begun to go beyond the adaptation of existing technology; today, many R&D units are also charged with the creation and renewal of core technological capabilities. Concurrently, as outlined in the chapter by Mowery, strategic intra-firm cooperation has taken on an added significance, especially in industries characterized by rapid technological development.

These new types of foreign R&D units are of two kinds. Some are established as companies diversify into new product areas and technologies through *foreign acquisitions*. In this case, the competence of acquired R&D laboratories often constitutes the very reason for the acquisition. Thus, contrary to the case of horizontal acquisitions—where the work of acquired R&D units tends to duplicate that already undertaken elsewhere in the corporation—companies usually have strong incentives to maintain and expand acquired technical activities.

A second type of R&D unit is set up in order to *'tap into' a foreign scientific infrastructure*. Such laboratories are devoted to research rather than development and are becoming increasingly common not only in technology-intensive industries, such as pharmaceuticals, electronics and biotechnology, but also in some engineering industries.

The cost of R&D varies among nations and over time and is sometimes an ancillary motive for locating research activities abroad. In Bangalore in India, for example, a graduate researcher, educated at ivory league universities in the US, costs less than one-tenth of what he would cost in Stockholm. One square meter of an advanced biotech laboratory also costs less than one tenth of the cost of a similar square meter in Western Europe or in the US. The availability of certain specialized biotech researchers in Bangalore is over one hundred times greater in Bangalore, with its conglomeration of knowledge activities in biotech and software, than in Sweden. Astra, a Swedish pharmaceutical company, established its recent laboratory in Bangalore, not in Göteborg, the home town of one of its major subsidiaries (Astra-Hässle) or in the greater Stockholm area, where its head office and major R&D facilities are located. This example points at cost differentials and the availability of R&D inputs, such as manpower and facilities, the conglomeration of competence, and the small size of the home economy, as driving forces behind internationalization of R&D.

Of great interest (for wealth creation and diffusion through R&D activities) is the question as to whether national governments by means of *subsidies* or similar means can influence foreign-owned companies to take up R&D. However, both theoretical arguments and the scant empirical evidence available [26, 28, 43] suggest that this is hardly the case. As argued by Rugman [48], R&D subsidies may have some influence on subsidiaries that already carry out R&D, but the uncertainty inherent in all political measures of this sort probably makes companies reluctant to make themselves dependent on them.

1.3.4 Company Determinants of Foreign R&D Shares

The relative importance of foreign R&D activities, as measured by e.g. the relative proportion of foreign R&D costs, varies greatly between companies and industries. A wide variety of firm-specific factors driving the internationalization of R&D have been suggested and analyzed (see [35] for a recent overview).

It is generally assumed that the mission for the bulk—or at least a large proportion—of foreign R&D is to support local production, be it through adaptation of products and processes to local conditions, through improvement of existing products or through new product development. It is therefore not surprising that the share of foreign R&D tends to correlate with the share of foreign production (and other measures of company internationalization) [27, 39, 41, 44].

Some results from previous research suggest an association between age, size and stage of corporate development, on the one hand, and degree of internationally performed R&D, on the other [19, 20, 25, 31, 41]. This indicates that the propensity to perform R&D abroad increases with the general growth and development of a company, not merely with its internationalization. Several supporting tendencies help to explain this trend.

Successful foreign subsidiaries tend, with time, to accumulate technical, managerial and financial resources. A local subsidiary management frequently has an inducement to allocate some of these resources to product development. Sometimes, the evolution of routine technical activities into design and development can take place 'spontaneously', i.e. without formal authorization from headquarters [19, 36, 37, 49]; sometimes, e.g. when such changes are seen to be a break with established policy, they require formal top management approval [46].

The probability that such an evolution—from routine technical activities associated with customization of products, technical service, quality control and manufacturing support into proper R&D—will take place depends on the level of technological sophistication of the activities. Hence, Hewitt [25] has shown that industries with a high percentage of 'professional, technical and kindred workers'—where foreign subsidiaries typically employ a high share of technical personnel—tend to perform larger proportions of their R&D abroad.

Learning effects probably contribute to this tendency. As pointed out by Hewitt, it takes time for a company to learn to appreciate the advantages of foreign R&D. Similarly, it takes time to develop the systems and procedures required to bring down the costs of coordination and control of a geographically decentralized R&D organization [39]. Increased international experience in marketing and production makes firms better informed on local R&D climates and differences among regions in terms of the quality of R&D resources. This, in turn, may lead to increased international competition.

A clear relationship has been found between foreign R&D and the international pattern of manufacturing. Hirschey and Caves [27] and Zejan [55] found a positive influence of foreign R&D and export shares on foreign subsidiary sales. Companies that concentrate manufacturing to a few specialized factories have greater possibilities of locating some R&D capacity there than companies with manufacturing dispersed over smaller factories in many countries. Foreign subsidiary exports to the United States varied negatively with the share of R&D abroad. However, in Pearce's [44] replication of the American study, using 1982 data, these relationships were not statistically significant.

The potential effects of this factor are especially relevant in Europe, where the creation of free trade areas, in combination with the reduction of transportation costs, has made concentration of production to specialized plants—and the assignment of so-called 'global product mandates'—an increasingly attractive alternative [13, 19, 16].

1.3.5 Home and host country impact

When analyzing implications of foreign R&D units for the host country, two questions must be considered; (1) What type of R&D is undertaken: engineering, development or research? and (2) What are the direct and indirect benefits of R&D undertaken by foreign firms relative to the costs (including alternative uses of host country R&D resources) of such inward investments?

The article by Dunning in this volume summarizes the current state of knowledge regarding the direct and indirect effects of MNCs' R&D activities on the innovative capability of the host country. There are two different perceptions underpinning different policies in the past. One view considers inward investments in R&D to be in general beneficial to economic growth and general welfare. Such investments, it is argued, provide technology and managerial skills which create indirect positive effects for the host country. These effects include technical support and assistance to local suppliers and customers, copying of foreign R&D units' output by local firms, contract jobs from foreign R&D units to local R&D organizations, etc. Inward foreign direct investment is seen to speed up the pace of economic growth and organizational restructuring by providing technology and managerial skills at a lower cost, through competitive stimulus and other spill-over effects. Examples of countries that appear to have developed their policies according to this thinking are Austria and Singapore.

The other view emphasizes that local R&D activities run by foreign firms tend to tap into unique local R&D resources with little or no benefit to the host country. Working on problems of little relevance to the local economy, they may represent little more than a disguised 'brain-drain', diverting scarce technical resources from more useful purposes. Thus, some observers, rightly or wrongly, have accused some European and especially Japanese MNCs of exploiting the rich scientific resources at major US universities without providing any tangible benefit. (Often, the argument is extended to include the virtual impossibility for US companies to exploit Japanese universities, due to their lack of front-line research, language barriers and numerous other obstacles.) In a similar vein, foreign acquisitions of local technology-based companies are often lamented as a mechanism by which vital technology, production and exports are diverted from their country of origin.

Figure 1.1 summarizes the main features of internationalization of R&D, its causes and effects, as they are commonly identified in the literature.

Figure 1.1 Overview of internationalization of R&D, its contexts and causality

1.4 PRESENT AND FUTURE OF THE RESEARCH AGENDA

1.4.1 Structure of the book

With some simplification, the research agenda from which the articles collected in this volume are drawn can be divided into three parts: The first involves the *'mapping out' of the territory*, i.e. the empirical determination of the relative volume, scope and geographical pattern of foreign R&D in multinational enterprises. The second involves the analysis of the *macro-economic effects and policy issues* of foreign R&D activities on both home and host countries. The third concerns the *micro-economic effects and managerial issues* confronting managers in MNCs with geographically and organizationally decentralized R&D, i.e. questions regarding the design of appropriate systems, procedures and mechanisms to ensure the attainment of efficiency, local adaptation and worldwide learning in an international R&D network.

The following chapters have been ordered roughly according to this logic. The second chapter by Dunning reviews a wide range of research and empirical sources to present a synthesis of major findings and issues in the field. The author analyzes global differences in technical strengths and relates these to observed patterns in the foreign R&D activities of multinational companies. In most countries, such activities represent the exception rather than the rule; the bulk of R&D is still performed in central home country-based laboratories. Nevertheless, foreign R&D volumes are far from negligible and Dunning proceeds to analyze the questions of home and host country impacts, drawing on recent experience in the UK.

Patel and Pavitt (Chapter 3) and Cantwell (Chapter 4) both use US patent data in order to investigate the extent and pattern of foreign R&D activities. Both articles draw on a database developed at the University of Reading and the Science Policy Research Unit (SPRU) at the University of Sussex. This makes it possible to study—for the first time—the output of foreign relative to home country based R&D activities. Patel and Pavitt emphasize the continued importance—in all but a few small European countries—of home-based R&D. With a slightly different angle, Cantwell explores the relationship between multinational R&D locations and national centers of innovative excellence in different industries.

Håkanson, in Chapter 5, analyzes the impact of different country characteristics as factors influencing the location of foreign R&D units in Swedish multinationals, demonstrating that the relative importance of different factors varies with the type of R&D performed.

Casson, Pearce and Singh (in Chapter 6) and Pearce and Singh (in Chapter 7) report survey results on the roles and management of foreign R&D laboratories. Like Cantwell, these authors emphasize the significance of international networking and explore some of the managerial implications

Introduction and Overview ─────────────────────────── 15

of this trend. Pearce and Singh find only limited support for the hypothesized importance of 'supply side' factors as motives for international decentralization of R&D, a result echoing the conclusions by Patel and Pavitt.

Drawing on case study evidence, De Meyer (in Chapter 8) discusses the role of planning and control systems as instruments for learning, relating his analyses to recent concepts in the literature on MNC organization.

Drawing on a questionnaire survey and interviews in a large number of multi-technology companies in Sweden, Japan and the US as well as on a case study of mobile telephone systems, Granstrand and Sjölander (Chapter 9) explore the implications of technological progress, internationalization of technology sourcing (acquisition) and technological diversification for the formulation and implementation of corporate strategies.

In Chapter 10 Mowery extends the discussion of international R&D management to one of the most significant alternative mechanisms for renewing firm-specific technological assets by tapping into foreign sources of technological and scientific know-how, i.e. cross-border interfirm R&D cooperation.

The internationalization of R&D and technology entails considerable challenges both for the managers responsible for the efficient exploitation of scarce but geographically dispersed technological resources and for policy makers in home and host countries charged with the development and implementation of efficient technological and industrial policies. However, we are as of yet only at the beginning of the road towards the answers to the many normative issues involved. Nevertheless, with the aid of the research undertaken so far, some of which is reported in this book, it is possible to define a number of crucial areas for future research. It should be remembered that here, as in most scientific endeavors, the problem and process of finding answers are not nearly as difficult as those of formulating the right questions.

1.5 REFERENCES

[1] L. Ahnström (1973), *Styrande och ledande verksamhet i Västeuropa*, EFI, Stockholm (in Swedish).
[2] W.B. Arthur (1988), Competing technologies: an overview. Published in Dosi et al. (eds), *Technical Change and Economic Theory*, Pinter Publishers, London.
[3] C.A. Bartlett (1986), Building and Managing the Transnational: The New Organizational Challenge, in M.E. Porter (ed.), *Competition in Global Industries*, 367–401. Boston, Harvard Business School Press, Boston, MA.
[4] C.A. Bartlett and S. Goshal (1989), *Managing Across Borders*, Harvard Business School Press, Boston, Ma.
[5] C.A. Bartlett, Y. Doz and G. Hedlund (eds) (1990), *Managing the Global Firm*, Routledge, London.
[6] S. Burenstam Linder (1961), *An Essay on Trade and Transformation*, Almqvist & Wiksell, Uppsala.

[7] M.C. Casson (1987), *The Firm and the Market*, Basil Blackwell, Oxford.
[8] A.J. Cordell (1971), *The Multinational Firm, Foreign Direct Investment and Canadian Science Policy*, Science Council of Canada, Special Study.
[9] A.J. Cordell (1973), Innovation, the Multinational Corporation: Some Implications for National Science Policy, *Long Range Planning* 6, 22–29.
[10] G. Dosi (1988), Sources, Procedures, and Microeconomic Effects of Innovation, *Journal of Economic Literature*, 26, 1120–1171.
[11] G. Dosi, C. Freeman, R. Nelson, G. Silverberg and L. Soete (eds) (1988) *Technical Change and Economic Theory*, Pinter, London.
[12] G. Dosi, K. Pavitt and L. Soete (1990), *The Economics of Technical Change and International Trade*, Harvester Wheatsheaf, New York.
[13] Y. Doz (1978), Managing Manufacturing Rationalization within Multinational Corporations, *Columbia Journal of World Business* 82–93.
[14] J.H. Dunning (1988), The Eclectic Paradigm of International Production: A Restatement and Some Possible Extensions, *Journal of International Business Studies* 19, 1–31.
[15] H. Ergas (1987), The Importance of Technology Policy, in P. Dasgupta & P. Stoneman (eds), *Economic Policy and Technological Performance*, 51–96, Cambridge University Press, Cambridge.
[16] T.M. Flaherty (1986), Coordinating International Manufacturing and Technology, in M.E. Porter (ed.), *Competition in Global Industries*, 83–109, Harvard Business School Press, Boston, MA.
[17] C. Freeman (1982), *The Economics of Industrial Innovation*, 2nd ed., MIT Press, Cambridge, MA.
[18] C. Freeman (ed.) (1990), *The Economics of Innovation*, Edward Elgar, Aldershot.
[19] O. Granstrand (1979), *Technology, Management and Markets. An Investigation of R&D and Innovation in Industrial Organizations*. Chalmers University of Technology, Göteborg, Sweden. (Published in abridged and revised form by Pinter Publishers, London, 1982.)
[20] O. Granstrand and I. Fernlund (1978), Coordination of Multinational R&D. A Swedish Case Study, *R&D Management*, 9, 1–7.
[21] O. Granstrand, C. Oskarsson, N. Sjöberg and S. Sjölander (1990), Business Strategies for New Technologies. Paper presented at the conference on 'Technology and Investment' arranged by IVA, OECD and the Swedish Ministry of Industry, 1990. Published in E. Deiaco et al. (eds) *Technology and Investment*, Pinter, London, pp. 64–98.
[22] G. Hedlund (1986), The Hypermodern MNC—A Heterarchy?, *Human Resource Management*, 25, 9–35.
[23] E. Helpman and P. Krugman (1989), *Trade Policy and Market Structure*, MIT Press, Cambridge.
[24] J.F. Hennart (1982), *A Theory of Multinational Enterprise*, University of Michigan Press, Ann Arbor.
[25] G. Hewitt (1980), Research and Development Performed Abroad by U.S. Manufacturing Multinationals, *Kyklos* 33, 308–327.
[26] G. Hewitt (1983), Research and Development Performed in Canada by American Manufacturing Multinationals, in A.M. Rugman (ed.), *Multinationals and Technology Transfer—The Canadian Experience*, Praeger, New York.
[27] R.C. Hirschey and R.E. Caves (1981), Research and Transfer of Technology by Multinational Enterprises, *Oxford Bulletin of Economics and Statistics* 43, 115–130.
[28] J.D. Howe and D.G. McFetridge (1976), The Determinants of R&D Expenditures, *Canadian Journal of Economics* 9, 57–71.

Introduction and Overview — 17

[29] S. Hymer (1976), *The International Operations of National Firms*, MIT Press, Cambridge, MA.

[30] L. Håkanson (1980), *Multinationella företag: FoU-verksamhet, tekniköverföring och företagstillväxt*, SIND 1980:4, Liber, Stockholm. (In Swedish.)

[31] L. Håkanson (1981), Organization and Evolution of Foreign R&D in Swedish Multinationals, *Geografiska annaler*, Ser. B **63**, 47–56.

[32] L. Håkanson (1983), R&D in Foreign-Owned Subsidiaries in Sweden, in W. Goldberg (ed.), *Governments and Multinationals*, Oelgeschlager, Gunn & Hain, Cambridge, Mass.

[33] L. Håkanson (1989), International Decentralization of R&D—The Organizational Challenges, in C. Bartlett, Y. Doz and G. Hedlund (eds), *Managing the Global Firm*, Addison-Wesley, London.

[34] L. Håkanson and R. Nobel (in press), Foreign Research and Development in Swedish Multinationals, Institute of International Business, Research Paper 89/3, *Research Policy*.

[35] L. Håkanson and R. Nobel (in press), Determinants of Foreign R&D in Swedish Multinationals, Institute of International Business, Research Paper 90/1, *Research Policy*.

[36] L. Håkanson and U. Zander (1986), *Managing International Research & Development*, Sveriges Mekanförbund, Stockholm.

[37] L. Håkanson and U. Zander (1988), International Management of R&D: The Swedish Experience, *R&D Management* **18**, 217–226.

[38] P. Krugman (ed.) (1986), *Strategic Trade Policy and the New International Economics*, MIT Press, Cambridge.

[39] S. Lall (1979), The International Allocation of Research Activity by US Multinationals, *Oxford Bulletin of Economics and Statistics* **41**, 313–331.

[40] E. Mansfield and A. Romeo (1980), Technology Transfer to Overseas Subsidiaries of U.S.-Based Firms, *Quarterly Journal of Economics* **95**, 737–750.

[41] E. Mansfield, D. Teece and A. Romeo (1979), Overseas Research and Development by U.S.-Based Firms, *Economica* **46**, 187–196.

[42] R. Nelson and S. Winter (1982), *An Evolutionary Theory of Economic Change*, Harvard University Press, Cambridge, MA.

[43] D.A. Ondrack (1983), Responses to Government Industrial Research Policy: A Comparison of Foreign-Owned and Canadian-Owned Firms, in W. Goldberg (ed.), *Governments and Multinationals*, Oelgeschlager, Gunn & Hain, Cambridge, Mass.

[44] R.D. Pearce (1989), *The Internationalization of Research and Development by Multinational Enterprises*, Macmillan, London.

[45] M.E. Porter (1990), *The Competitive Advantage of Nations*, Macmillan, London.

[46] R.C. Ronstadt (1976), International R&D: The Establishment and Evolution of Research and Development Abroad by Seven U.S. Multinationals, *Journal of International Business Studies* **9**, 7–24.

[47] R.C. Ronstadt (1977), *Research and Development Abroad by U.S. Multinationals*, Praeger, New York.

[48] A.M. Rugman (1981), Research and Development by Multinational and Domestic Firms in Canada, *Canadian Public Policy* **7**, 604–616.

[49] S. Sjölander (1985), *Management of Innovation*, Department of Industrial Management and Economics, Chalmers University of Technology, Göteborg, Sweden.

[50] L. Soete (1991), *Trade and Technology Policies and International Competitiveness*. Paper presented at the Chalmers/IVA symposium on Economics of Technology, 1991, available through the Department of Industrial Management and Economics, Chalmers University of Technology, Göteborg, Sweden.

[51] L.W. Steele (1975), *Innovation in Big Business*, Elsevier, New York.
[52] R. Vernon (1966), International Investment and International Trade in the Product Cycle, *Quarterly Journal of Economics* **80**, 190–207.
[53] R. Vernon (1974), The Location of Economic Activity, in J.H. Dunning (ed.), *Economic Analysis and the Multinational Enterprise*, Allen & Unwin, London.
[54] M. Wortmann (1990), Multinationals and the Internationalization of R&D: New Developments in German Companies, *Research Policy* **19**, 175–183.
[55] M.C. Zejan (1988), *Studies in the Behavior of Swedish Multinationals*, Ekonomiska studier utgivna av nationalekonomiska institutionen, Handelshögskolan vid Göteborgs universitet, No. 23. (In Swedish.)
[56] J. Zysman, S. Cohen, G. Dosi and L. Tyson (1990), Trade, technology and national competition. Published in E. Deiaco et al. (eds) *Technology and Investment*, Pinter Publishers, London.

Chapter 2

Multinational Enterprises and the Globalization of Innovatory Capacity

JOHN H. DUNNING

This chapter first describes the extent to which research and development (R&D) is being decentralized within multinational networks. It then goes on to review the impact of international direct investment on the innovatory capacity of host countries, and the role which government policy may play in influencing multinational enterprises (MNEs) in their international R&D strategies, The chapter concludes with some speculations about the future form and location of MNE high value activity. In particular, it suggests that there will be a greater amount of intra-firm, intra-industry trade in R&D, and the emergence of more pluralistic modes of the cross-border management of technological and organizational capacity.

2.1 INTRODUCTION

One of the most frequently asked questions about the impact of multinational enterprises (MNEs)[1] on both host developing and developed countries is 'What is their contribution to the development of indigenous innovatory capability?' Increasingly, countries are judging their economic success and strategic security by their ability to create, upgrade, and sustain the human and physical assets necessary to produce wealth. Given the availability of

---

[1] Defined as enterprises which control and coordinate value added activities across national boundaries.

Technology Management and International Business: Internationalization of R&D and Technology.
Edited by O. Granstrand, L. Håkanson and S. Sjölander. © 1992 John Wiley & Sons Ltd

natural resources, the capability to efficiently convert these resources into existing goods and services, and to innovate new goods and services is now center stage. So, indeed, is the entrepreneurial ethos and vision of the potential innovators and their suppliers, competitors, and customers, but this is not the primary subject of this chapter.

The focus of this contribution is on the extent to which MNEs can and do act as cross-border organizers and diffusers of innovatory capabilities, and so further the globalization of technological capacity. We shall argue that such capacity may be measured in terms of both the stock of human and physical capital, or by some measure of the output of such assets, e.g. patented innovations and improvements in production and/or organizational techniques. We shall be primarily concerned with the ownership and location of innovatory assets, and, among the questions specifically addressed are the following.

What role do MNEs play in the generation of technological assets? Is the location of research and development (R&D) becoming more or less centralized? What kinds of R&D do MNEs delegate to their affiliates? What changes are occurring in the ownership or organization of international production? Does the presence of foreign-owned R&D facilities stimulate or inhibit the development of local technological or absorptive capabilities? Under what circumstances will MNEs promote a more efficient international division of labor of the production of new technology? What should be the policy responses of home and host governments to the innovatory strategies of MNEs, and to their impact on national economic goals?

2.2 INNOVATORY CAPACITY BY COUNTRY: THE LOCATION OF CAPACITY

Some statistics on R&D expenditures of non-Communist countries are set out in Table 2.1. All the usual cautions apply to these data. They are fragmentary and rarely directly comparable. Nevertheless, the general picture they portray is a fairly clear one. In 1986/7, 82% of the world's expenditure on R&D and 69% of the personnel engaged in R&D were accounted for by five countries, viz the US, Japan, France, the UK and West Germany; and 91% and 84% by ten countries[2]. Only 4.3% of R&D expenditure was undertaken by developing countries; indeed, their combined spending on R&D in the mid-1980s (about $10 billion at 1982 prices) was less than that of France in 1987 ($13.7 billion at 1982 prices). However, because of the greater labor intensity of these R&D activities, developing countries were responsible for a higher proportion—viz 12.6%—of the personnel employed in such activities[3].

---

[2] The above five and Canada, Italy, the Netherlands, Sweden, and Switzerland.
[3] In 1986, for example, India employed 172370 people in R&D activities; and was ranked sixth as a global employer.

Table 2.1 Geographical distribution of R&D expenditure in constant (1982) $ billions, 1986–1987

	$b.	%	$b.	%
Developed Countries				
North America			105.6	46.3
of which:				
US	100.8	44.2		
Western Europe			71.1	31.2
of which:				
West Germany	19.4	8.5		
United Kingdom	13.8	6.1		
France	13.7	6.0		
Italy	7.4	3.2		
Sweden	4.0	1.8		
Japan			39.1	17.2
Other developed countries			2.2	1.0
Developing Countries[a]			9.9	4.3
of which:				
India	1.5	0.7		
Brazil	1.4	0.6		
South Korea	1.3	0.6		
Argentina	1.1	0.5		
All Countries			227.9	100.0

Source: National Science Foundation [28].
[a] Including Yugoslavia.

Output data present a broadly similar picture[4]. Of all the patents registered in the US in the period 1970–1988, 90% were accounted for by firms which are located in the US, Japan, France, the UK, and West Germany; and less than 1% were from developing countries. Of the developing countries, Taiwan, South Korea, Brazil, Mexico, Hong Kong and India are currently among the leading filers of patents.

The evidence suggests that there has been some geographical dispersion of the location of innovatory capacity since 1970 [45, 30]. In that year, for example, the USSR, Canada, and the US accounted for 65.4% of world R&D expenditure and 57.4% of the scientists and engineers engaged in R&D. By 1980, these proportions had fallen to 47.7% and 54.6%; whereas in 1970, 75% of the patents registered in the US were granted to North American-owned firms; in 1985, this proportion had fallen to 57.0%.

The most noticeable convergence in innovatory capacity has, however, occurred among the five leading industrial countries, i.e. the US, France, West

---

[4] A rank correlation between the R&D expenditure (1979–1986) and registered patents (1970–1985) of the ten largest country spenders on R&D was 0.794 [2].

Germany, the UK, and Japan. As Table 2.2 shows, in 1970 the US accounted for 61.7% of the total R&D expenditure of these countries, employed 59.8% of the scientists and engineers engaged in R&D, and was responsible for 80.0% of the patents registered in the US. By 1987, the respective percentages had fallen to 54.0%, 50.9% and 58.5%. Over the same period, the corresponding percentages accounted for by the three leading European innovating nations had risen from 26.1%, 21.3% and 15.4% to 25.1%, 23.0%, and 18.1%; and those of Japan from 12.3%, 18.9%, and 4.5% to 20.9%, 26.1%, and 23.3%.

UNESCO data suggest that there was also a noticeable increase in the technological capacity of developing countries. Their contribution to world R&D expenditure rose from 2.5% to 6.2% between 1970 and 1987, and that of R&D scientists and engineers from 8.5% to 11.2% [45]. They more than doubled their share of world patenting between 1963–70 and 1974–84. Some newly industrializing countries are also gaining an increasing share of new patents registered in some (mainly fabricating) industries.

In spite of some convergence in the innovating capabilities among the leading industrial nations, the sectoral pattern of research expenditure and patent performance differs considerably between countries[5]. Tables 2.3 and 2.4 set out some details. They show, for example, that, in 1986, the US had a research bias in aerospace and instruments, while, between 1978 and 1984, she had a particularly strong revealed patenting advantage[6] in aircraft and (surprisingly perhaps) in textiles, clothing and leather goods. Japan's R&D

Table 2.2 Distribution of (indicators) of innovating capability between five leading innovating countries 1970–1987

	R&D expenditure constant 1982 $ (billion)				R&D personnel (thousands)				Patents			
	1970		1987		1970		1987		1970		1987	
		%		%		%		%		%		%
US	62.4	61.7	100.8	54.0	543.8	59.8	791.1	50.9	47,077	80.0	40,496	58.5
Japan	12.4	12.3	39.1	20.9	172.0	18.9	405.6	26.1	2,625	4.5	16,158	23.3
West Germany	9.9	9.8	19.4	10.4	82.5	9.1	151.5	9.7	4,435	7.5	7,307	10.6
France	7.1	7.0	13.7	7.3	58.5	6.4	108.2	7.0	1,731	2.9	2,661	3.8
UK	9.4[a]	9.3	13.8	7.4	52.8[a]	5.8	98.7[b]	6.3	2,954	5.0	2,583	3.7
	101.2	100.0	186.8	100.0	909.6	100.0	1555.1	100.0	58,822	100.0	69,205	100.7

Source: National Science Foundation [28].
1969[a].
1986[b].

---

[5] For recent discussions on this topic, see especially [2, 37].
[6] See Table 2.4 for definition.

MNEs and the Globalization of Innovatory Capacity ——————— 23

Table 2.3 Distribution of privately funded R&D expenditures by manufacturing sectors, 1986

Industries	US	Japan	West Germany	France	UK
Mainly Processing Industries					
Food & allied products	1	2.6	1.1	1.4	2.6
Pharmaceuticals	4.7	6.2	22.1	7.6	9.9
Other chemicals	6.8	10.7		10.4	9.9
Petroleum refining	1	1.2	0.7	3.0	1.2
Rubber and plastics	1.5	2.6	1.7	2.6	0.8
Textiles	a	1.1	0.5	0.6	0.5
Metals	0.6	6.1	2.1	2.2	1.2
	13.6	30.5	28.2	27.8	26.1
Mainly Fabricating Industries					
Aerospace	23.4	neg	7.2	20.3	17.8
Non-electrical machinery	14.4	12.7	14.5	9.4	13.0
Electrical and electronic products	22.7	29.1	27.2	23.6	31.1
Instruments	7.2	3.6	1.7	1.5	1.2
Transportation equipment	9.4	16.8	16.3	11.6	7.8
Fabricated metals	0.8	1.9	2.7	1.0	0.3
	77.9	64.1	69.6	67.4	71.2
Other	8.5	5.9	2.2	4.8	2.7
Total	100.0	100.0	100.0	100.0	100.0

Source: IMD/World Economic Forum Report [22] Table 9.11, p. 166/8.
a Included in Other.

strength and patenting performance appears to be especially noticeable in consumer electronics, office equipment, semi-conductors and motor vehicles. The UK's prowess lies in tobacco processing, chemicals (especially) pharmaceuticals and aircraft, and that of Germany in nuclear reactors and a wide range of chemicals. Italy's comparative advantage was most pronounced in household appliances and agricultural chemicals; and the Netherlands' in consumer electronics and food products. Developing countries (as a group) appear to have a revealed patenting advantage in food products, inorganic chemicals, and coal and petroleum products[7].

---

[7] It is interesting to note that, although the activity classification is not completely comparable, there appears to be a fairly close association between the comparative research strengths of a country and its revealed patenting advantages.

Table 2.4 Revealed patenting advantage of selected countries in selected industries 1978–84

Industries	USA	Japan	Germany (FDR)	UK	France	Others[a]
Food	1.16	0.65	0.69	0.99	1.02	1.89
Drink	1.00	0.69	1.22	0.75	0.76	0.60
Tobacco	1.02	0.22	1.27	2.25	1.42	0.78
Dyestuffs	0.91	0.72	1.37	0.99	1.02	0.58
Pharmaceuticals	0.79	0.66	0.97	1.85	1.43	0.73
Fabricated metal prod.	0.95	0.66	1.08	1.05	1.00	1.54
Mech. eng. (general)	1.01	0.69	1.16	1.03	0.95	1.28
Elect. equip. (general)	1.03	1.15	0.89	0.95	1.15	0.91
Semiconductors	0.96	1.59	0.81	0.69	0.93	0.50
Office equipment & communications	0.98	1.76	0.59	0.69	0.82	0.41
Radio & TV receivers	0.88	1.74	0.55	0.91	0.84	0.36
Motor vehicles	0.91	1.10	1.19	1.02	1.91	0.83
Aircraft	1.22	0.18	1.00	2.03	3.46	1.43
Textiles, clothing, etc.	1.22	0.41	0.75	0.87	1.41	2.35
Instruments	0.93	1.35	0.93	0.82	0.78	0.78

Source: Data compiled at the University of Reading using information on patent courts obtained through the Science Policy Research Unit, University of Sussex. The data on patenting were originally prepared by the Office of Technology Assessment and Forecast, US Patent and Trademark Office, with the support of the Science Indicators Units, US National Science Foundation.

[a] All countries, apart from those countries identified and Switzerland, Netherlands, Sweden, Italy and Denmark.

Revealed Patenting Advantage is defined as

$$RPA_{ij} = \frac{P_{ij} / \ddot{E}_i P_{ij}}{\ddot{E}_j P_{ij} / \ddot{E}_i \ddot{E}_j P_{ij}}$$ Where j = industry and i = country

It represents the share of world patenting (in the US) of a particular country in a particular industry divided by its share of world patenting (in the US) in all industries, over the period 1978/84.

Table 2.5 presents some details on the subject area of students enrolled for graduate courses in colleges and universities in selected countries. The data suggest that, relative to other nations, the US, Mexico and Korea place particular emphasis on business and organizational studies, while Japan and West Germany have a 'revealed' educational advantage in engineering and in arts and humanities. Mexico and Egypt are particularly strong in medical science, the UK, Canada and Korea in educational studies, India in arts and the humanities and West Germany in law and social sciences[8]. It is to be noted that these statistics take no account of 'on the job' training. Here the data strongly suggest that German, Japanese and Swiss firms have

---

[8] Other data suggest that there is a strong resistance of Japanese companies to hiring university trained PhDs. In engineering, for example, whereas in 1986, Japan produced about the same number of first degree graduates in engineering as did the US, it produced only 588 doctoral graduates compared to 3376 in the US [see 47].

MNEs and the Globalization of Innovatory Capacity

Table 2.5   Percentage distribution by field of study of graduate students[1] in selected countries

Subject	Developed countries					Developing countries			
	Canada (1988)	US (1985) (1987)	W. Germany[2] (1987) (1988)	UK (1987)	Japan (1983)	Mexico (1988)	India (1988)	Rep. of Korea (1988)	Egypt (1988)
Arts & humanities	12.5	7.8	17.4	11.5	13.0	2.5	50.1	15.4	11.9
Education	15.2	9.8	4.6	19.2	5.7	6.8	1.8	16.7	14.0
Law & social sciences	12.3	11.2	29.3	26.6	9.6	12.4	1.2	8.3	14.5
Business studies	15.1	23.2	1.8			24.6	16.2	19.2	13.9
Natural sciences & mathematics	10.7	8.8	13.9	17.4	10.5	10.5	18.4	8.0	6.8
Medical science & health related	13.2	10.0	7.6	6.4	16.3	29.1	4.6	9.2	18.0
Engineering	9.6	5.5	17.4	11.1	35.7	7.8	3.5	13.2	9.4
Other	8.9	23.7	8.0	7.8	9.2	6.3	4.2	10.2	11.5
	100.0	100.0	100.0	100.0	100.0	100.0	100.0	100.0	100.0

Source: UNESCO (1987) Table 3.14 and UNESCO (1990) Table 3.13. See [45].
[1] Programmes leading to a graduate university degree or its equivalent.
[2] Including programmes leading to first university degree or its equivalent.

allocated considerably more resources, particularly in the areas of their comparative innovatory advantages, than their counterparts in the UK, France and Italy[9].

2.3 INNOVATORY CAPACITY BY FIRMS: THE OWNERSHIP OF CAPACITY

There are no comprehensive data on the innovating capacity of firms by country of ownership. It is, however, known that, except for some European and a few US MNEs, the average proportion of R&D activity undertaken outside their home countries is quite small, and that, in the case of almost all Japanese MNEs, it is negligible. Moreover, the majority of foreign direct

---
[9] In a survey conducted by IMD and the World Economic Forum [22], a group of international business executives were asked to rank (on a scale of 0 = inadequate to 100 = adequate) the vocational training facilities offered by some 32 countries. Of these, Germany was ranked highest with a score of 77.2, followed by Japan (73.9) and Switzerland (70.8). Among the lowest scorers (among developed countries) were the UK (35.7), Italy (35.1) and France (43.9). According to Tom Peters (quoted in [28]), Japanese companies, on average, spend three times the amount on worker training programs than do US companies. No less significant is the fact that Japanese companies place more emphasis on training their workers to develop a broad range of skills which can be adapted to variations in the tasks required of them. By contrast, US training is much more sector or skill specific. Moreover, 57% of the PhDs in natural sciences and engineering in Japan were earned by industrial researchers submitting papers based upon their work for companies, rather than as a result of any formal training at Japanese universities [47].

investment (FDI) originates from countries (the Soviet Union apart) which spend the most on R&D, register the largest number of patents and record the highest enrollment in higher education[10]. This would suggest that the geographical distribution of innovatory capacity by country of location and by firms should be broadly similar. Later paragraphs will question this view. Moreover, even if this were the case, it would not be correct to assume that strategies and policies of MNEs have little or no effect on the innovatory capacities of the countries in which they operate. Far from it; Section 2.5 will show that, in a variety of ways, MNE activity can and does affect the innovatory capacity of the countries in which it operates[11].

Data compiled on 792 of the world's largest industrial companies suggest that in 1982 about 30% of their production was undertaken outside their home countries, but only about 12% of their R&D expenditures [14]. For US MNEs, about which the data are the most comprehensive, the corresponding figures were 26% and 9% [36]; and for the leading European multinationals 37% and 23%. In 1987, some 22.8% of the R&D expenditure by 20 of the largest Swedish MNEs in the engineering and chemicals industries was undertaken outside Sweden, a slight increase in the 20.6% recorded by these same firms in 1980 [19]. In 1980, Swiss industry spent 2.8 billion francs in Switzerland and 1.7 billion francs abroad on R&D activities. Over the previous decade, R&D expenditure had risen by 19% in Switzerland and 50% in other countries [41]. Between 1966 and 1982, the proportion of R&D expenditure of US MNEs accounted for by their foreign affiliates increased slightly from 6.5% to 8.8%[12]. Taken as a whole then, the foreign affiliates of MNEs would appear to be primarily producers of goods and services rather than of innovations.

However, these data conceal important differences between countries and industries. For example, while, in 1982, 95% of foreign R&D expenditure of US firms was concentrated in developed countries, only 75% of their output of goods was. In that same year, the R&D intensity of the value added activities of US affiliates[13] was well above average in Germany and the UK (which,

---

[10] In 1985, 72% of the foreign direct investment stock was accounted for by the US, UK, Japan and Germany, which were also responsible for 55% of the world's R&D expenditure.

[11] Various surveys have attempted to assess the extent to which foreign affiliates engage in R&D in the countries in which they operate. In the 1950s, Dunning [12] found that 75% of American manufacturing subsidiaries in the UK undertook some R&D. In the 1960s, the proportion was 45% in the case of American affiliates in Australia [3], and 48% of foreign affiliates in Canada [39]. Other data reviewed by Pearce [36] suggest that the corresponding figures (for either US or all foreign affiliates) at various dates in the late 1960s and 1970s, ranged from 25% in the case of Brazil, 35% in Mexico, and 39% in Belgium, to 60% in Australia (in the late 1970s).

[12] Since this chapter was completed, data for 1989 have become available. These suggest that the share of foreign based R&D expenditure has increased further to 9.0% (10.1% of R&D performed by and for the affiliates, or by and for the MNEs themselves).

[13] Measured as the percentage of the sales of these companies accounted for by R&D expenditure. In the UK it is 1.53, and in Germany it is 2.30.

MNEs and the Globalization of Innovatory Capacity — 27

between them, accounted for one-half of all R&D undertaken by US manufacturing affiliates), but zero or very low in most developing countries[14]. Around an average share of global R&D expenditure accounted for by foreign affiliates of 8.8%, the ratio varied between 28.2% in tobacco, 15.3% in pharmaceuticals, 14.8% in household appliances, and 16.4% in food and kindred products, to 5.6% in textile products, 5.2% in machinery, and 4% in primary and fabricated metals.

Moreover, the industrial structure of foreign direct investment is often very different from that of the countries in which they operate. Because of the advantages which MNEs have in creating and coordinating innovatory assets, they tend to be more concentrated in R&D intensive sectors of the host countries. However, in most cases, the propensity of foreign-owned affiliates to engage in R&D is about the same as, or less than, that of their indigenous competitors [25, 36]. One exception is the Republic of Korea where, in 1980, foreign firms spent 1.8% of their sales on R&D compared with 0.47% for Korean firms. Another is Belgium where, in 1976, US subsidiaries allocated 2.2% of their sales to R&D compared with 1.8% for Belgian uninational firms [46]. Additionally, in recent years, there has been an increasing tendency for R&D intensive activities to be cross-traded between industrialized countries, and especially between the larger European countries and the US[15].

For these reasons, it is not surprising that, viewed from a host country perspective, the R&D activities of multinational affiliates are often relatively quite important. Again, the data are fragmentary, but in Australia, Belgium, Canada, the UK, West Germany, South Korea, Singapore and India in the 1980s (to give some examples), the share of national R&D expenditure accounted for by foreign-owned firms exceeded 15%.

Data on the patents registered in the US by 727 of the world's largest firms are even more revealing. They show that a variable, but not insignificant, proportion of patents of the leading MNEs originate from discoveries made outside their country of origin; and that the degree of internationalization varies according both to the home country of the investing firm and the host country of the foreign affiliates.

Further data are provided by John Cantwell in his contribution to this volume[16]. *Inter alia*, these show that, around an average of 10.6% in 1938—86, the share of US registered patents attributable to research in locations outside the home country of the parent company, was as high as 71.3% for Belgium and 70.0% for Dutch firms, and as low as 1.2% for Japanese firms.

---

[14] Brazil has the highest research intensity ratio of 0.47.
[15] Part of this increase in intra-industry R&D activity is accounted for by the substantial growth in cross-border acquisitions and mergers over the past ten years. Indeed, an important *raison d'être* for some of these A&Ms has been to acquire new innovatory capacity.
[16] See Section 2.5, Tables 2.1–2.3.

Of the larger industrialized countries, UK firms recorded the highest dependence on the research output of their foreign affiliates of 44.9%. In seven of the eleven countries identified, and most noticeably in the US, there had been an increase in the patents attributable to the foreign affiliates of MNEs over the previous two decades. For Japan, however, the data suggest that an overwhelming increase in the rising R&D expenditure of Japanese firms has been undertaken in their home countries.

Part of the above differences reflect the different industrial composition of R&D spending by countries. John Cantwell's research also suggests that the proportion of US registered patents attributable to the foreign affiliates of firms varies noticeably between industries. It is generally lower among the higher research intensive sectors than it is in the medium to low research intensive sectors. And it is in these latter sectors that countries whose MNEs record the highest propensity to patent from their foreign affiliates also tend to be those with the highest revealed comparative advantage in R&D expenditure or patenting.

It is also worth observing that, over time, the share of patenting accounted for by the foreign subsidiaries of MNEs has shifted between sectors. In the mechanical engineering, office equipment (including computers), motor vehicles, food products, and other manufacturing sectors, the share has increased between 1969–72 and 1983–86, but, in others, it exhibited little change or has fallen. Broadly speaking, these movements have corresponded with the extent to which there has been a decentralization of R&D expenditure; but, perhaps, more interestingly, with the degree to which the multinationalization of firms and the cross-border rationalization of their value added activities has increased[17].

Finally, there is some evidence that non-resident-owned research activities tend to be drawn to those sectors in which the recipient country has a comparative patenting advantage. In the UK, for example, between 1978 and 1986, the average share of US patents of the largest non-UK firms attributable to innovations originating in the UK, as a proportion of all non-UK patenting due to research in all foreign locations, which was directed to sectors with a revealed technological advantage of greater than 1, was 17.4%. This compared with an average share directed to sectors in which the revealed technological advantage was less than 1, of 12.1%[18].

The main conclusion to be drawn from the statistical evidence on innovatory capability—sparse as it is—is that MNEs not only undertake the bulk of R&D in the world economy, but they affect the way in which it is owned, organized and located. It would also seem that although there has been some

---

[17] For further details see Section 2.5, Tables 2.1 and 2.2.
[18] Around an average of 14.3%. The data were derived from Tables 5.5 and 5.19 of Cantwell and Hodson [7]. All averages calculated were unweighted averages.

MNEs and the Globalization of Innovatory Capacity _____ 29

diffusion of innovatory capacity over the past two decades—particularly in some sectors and among the leading industrial nations—the great bulk of R&D activity continues to be concentrated in the home country of MNEs. Unfortunately, from the published data, it is not possible to separate the R&D expenditure incurred at different stages of the process; though a variety of surveys have suggested that a higher proportion of applied research or developmental activity is likely to be decentralized than that of fundamental or basic research.

2.4 WHY DO MNEs ENGAGE IN FOREIGN INNOVATORY ACTIVITIES?

R&D expenditures represent a particular form of value added by firms. While these expenditures are usually perceived to precede the manufacturing of a product, sequential adaptations and improvements both to the product and to the production process make it desirable to link innovatory activities with both past and future output.

Firms will normally engage in foreign value adding activities whenever they perceive that they possess certain technological, managerial, or organizational ownership specific advantages over their competitors, which are best exploited internally from a foreign location[19]. But, firms may also invest abroad to acquire advantages, which, coupled with those they already possess, will help them to maintain or improve their competitive positions. Firms choosing to engage in foreign innovatory activities may do so for several reasons. In particular, we might identify four main types of R&D[20].

(1) *Product, material or process adaptations or improvements*
These activities are made necessary by different product needs and/or production of marketing conditions in the foreign (or the home) country. They may be resource based, market oriented, or rationalized. They require many different kinds of skills and expertise, from designers to applied scientists, engineers and technicians, but rarely the research scientist. External contact is mainly with suppliers of intermediate products and with final customers. The foreign R&D facilities essentially supply support services to those of their parent companies; and their success primarily depends upon, and affects, the 'know-how' rather than the 'know why'

---

[19] For recent reviews of the literature on the theory of international production, see [13] and [7].
[20] For alternative classification of R&D and a survey of the literature on this subject, see [36].

capabilities of the country in which they are located[21]. Occasionally, country-specific differences in resource endowments and capabilities, or in host government policies, may force a complete reappraisal of the affiliates' production methods and/or products; in which case, R&D becomes of type (2).

Much of the R&D undertaken by MNEs outside their home countries is of this first kind; for example, 57% of a sample of 218 Japanese MNEs questioned in 1990 indicated that, according to the Export-Import Bank of Japan, the main objective of their foreign R&D facilities was to develop products specifically tailored to the need of local customers. Moreover, except in a few sectors, notably mining, food processing and pharmaceuticals, and in a few countries, e.g. Brazil, India, South Korea, and Mexico, it is the only R&D carried out in developing countries. However, in some cases, the R&D consequent upon MNE activity is externalized to upstream sellers or downstream buyers. We shall take up this point in Section 2.5.2.

(2) *Basic materials or product research*

This is usually undertaken for two main reasons. The first is that inputs for this kind of research are immobile, e.g. tea plantations, bauxite mining, a particular climate or ecological condition, quality improvement techniques for agribusiness, etc. The second is that the need for continual testing and interaction with customer requirements necessitates a local R&D facility. The latter requirement is similar to that of the close proximity needed between R&D and production in the first stage of the product cycle. The output of this research may be used for products sold to both local and export markets.

This second type of R&D is likely to make more demands on the local innovatory infrastructure. It is more of a 'know why' than a 'know how' variety; and thus is more likely to flourish where it is part of a cluster of similar R&D activities and/or has access to University or cooperative research institutions.

(3) *Rationalized research*

This research has its equivalent in rationalized or cost minimizing production. To gain economies of scale and scope, MNEs may choose to concentrate particular kinds of research in particular foreign countries, the output of which they export to other parts of their network of activities. Like the second type of R&D, specialized or rationalized research usually requires a sophisticated innovatory infrastructure.

(4) *To acquire or gain an insight into foreign innovating activities*

As the ownership of R&D becomes increasingly concentrated, but its location becomes more dispersed—at least within the main industrialized countries—many MNEs, particularly in the technologically intensive

---

[21] For a discussion of the differences between 'know how' and 'know why', see [26].

MNEs and the Globalization of Innovatory Capacity 31

sectors, are finding it desirable to have an R&D, as well as a manufacturing presence, in the main innovating centers. Competitive pressures and the escalating costs of R&D are also leading an increasing number of companies to conclude cross-border research alliances[22]. Companies like IBM, Philips, ICI and Sony all have R&D facilities in Europe, the US and Japan. Countries anxious to attract high value activities of firms are attempting to create centers of innovatory excellence[23]. Some MNEs from developing countries are also investing in Europe and North America to acquire innovating capabilities in the same way as firms from industrialized countries invest in some developing countries to acquire raw materials or low cost labor.

The extent to which R&D activities are undertaken by MNE foreign affiliates will depend on, first, whether they are undertaken at all, and second on the relative advantages offered by a domestic and foreign location. Unlike foreign production, which, in most cases, requires some kind of competitive advantage on the part of the investing firm, research is undertaken to create or acquire an advantage. However, new R&D activity is not independent of the stock of technological capacity (including the learning experience) owned by the firm, and/or the availability of complementary assets. To the extent that the possession of these complementary assets may lower the marginal costs of R&D, the MNE (for all the usual reasons) may have an advantage in seeking out, monitoring and incorporating the research output of other firms into its own research portfolio. Hence, the desire of US multinationals to have a research presence in Europe, and of European and Japanese multinationals to have a research base in the US.

Finally, it is worth recalling that the market for innovatory resources is generally considered highly imperfect. When such imperfections lead to the price of such resources in foreign countries being relatively lower

---

[22] Noticeable recent examples include the collaboration between Siemens and IBM (in 1990) to jointly develop a memory chip capable of storing 64+million bits of information; that between Ciba Geigy and Tanox Biosystems (1989) to develop a therapeutic agent against AIDS; that between Ciba Geigy and Nestlé (in 1988) to undertake basic research in microbial genetics; that between Aerospatial and British Aerospace (in 1990) for the development of a second generation of supersonic airliners; and that between Thomson and Philips (in 1990) to develop a European system of high definition television (HDTV). For a recent examination of cross-border technological strategic alliances, see [18].
[23] A recent example is the setting up of Tsukuba Science City in the Ibaragi Prefecture of Japan as an international center for R&D. Already, several government research industries and colleges are concentrated at Tsukuba; and at least eleven foreign chemical companies have established R&D laboratories. According to the President of one of these companies 'Bayer', the importance of a corporate R&D triad stretching between Japan, the US and Europe is an essential part of a successful international strategy of any large MNE[24]. Another example of a major R&D investment in Japan is that of DuPont Electronics, which has committed $128 million in a facility at Yokohama, designed *inter alia* to afford the US company a window into the latest technological advances in electronic componentry in Japan.

(in relation to their opportunity and replacement cost), than in home countries, the MNE may have an additional reason for engaging in foreign-based R&D[24].

2.5 SOME LOCATIONAL PATTERNS

We have already indicated that there are very few comprehensive data on the location of innovatory activities by MNEs. Such evidence as we do have, including that obtained in a research project recently completed at the University of Reading [8], suggests that the foreign affiliates of US-based MNEs are among the most likely to engage in rationalized research; and particularly so in the pharmaceuticals, computer electronics, motor vehicles, and industrial instruments sectors. Basic research, undertaken outside their home countries, is more common in European MNEs in the resource-based processing industries. But most firms, and especially those which are the least experienced as foreign investors, only undertake R&D of type (1). However, we suspect that the fastest growing foreign R&D activities are those which are a combination of types (2) and (4). This is particularly likely to be so in the case of the R&D expenditures of foreign-owned firms in the US, which, between 1977 and 1985, rose by 5.6 times compared with an increase of sales of 3.2 times [44].

Let us, however, return to consider the firm specific characteristics influencing the organization and location of research. First, the organization of research generally favors large firms. This is partly because R&D often involves the integration of several different technologies and/or resource inputs, which are best coordinated under the same governance; and partly because R&D is often very expensive. This is not to deny that small firms may sometimes be important innovators. But the evidence suggests that, in technology intensive sectors, at least, the commercialization of innovations is becoming increasingly complex, costly, and risky; and it is the ability to bear these costs and risks, and to organize this phase of the R&D process, which the larger MNEs are particularly well equipped to do. They are also more likely to be able to cope with the ever decreasing time span between the discovery and obsolescence of innovations than are smaller firms.

We have suggested that, in recent years, private collaborative research and development has been one of the ways in which firms have reacted to the rising costs of R&D. This has led to an expansion in cross-border arrangements, particularly among developed country firms. For example,

---

[24] It is claimed, for example, that this is particularly the case in the UK, which results, in the words of one scholar [40] 'in the use of (technological) resources by a MNE representing a subsidy to the research of the MNE by the UK taxpayer'.

MNEs and the Globalization of Innovatory Capacity ——————— 33

between 1974 and 1985 Ghemawat, Porter and Rawlinson [16] traced 423 cross-border coalitions involving technology development. A later study of some 1155 cross-border cooperative agreements in the information technology sectors between 1980 and 1986 reveals that 27% were innovatory-related [4]. Other research by Alic [1] and Chesnais [9] give examples of pre-competitive R&D alliances of cooperative ventures between firms in the private sector and of jointly funded government-industry ventures in pre-competitive R&D, downstream technology development, customer-supplier agreements and co-production technology based agreements linking direct competitors.

If the organization of R&D favors large firms, the logistics of locational needs generally make for a concentration of R&D facilities in those countries with a good educational, innovatory and communications infrastructure. But what of the recent changes affecting location of R&D by multinationals? First, consider the amount of adaptive research. Here the evidence is mixed. On the one hand, firms seem to be paying more attention to a decentralization of research. On the other, any movement towards a world product mandate and the employment of production methods which are technology intensive, and use comparatively few factor endowments, is likely to have the opposite effect.

As firms increase their degree of multinationality, and as the supply capabilities of countries become more widely dispersed, it might be predicted that this might lead to more geographical specialization of R&D. The opposing view is that the increasing technological and educational infrastructure needed for R&D points to a greater clustering of innovatory activities in complementary sectors to gain agglomerative economies. Our reading of contemporary trends is that, while many of the leading MNEs are seeking to establish or acquire R&D facilities in the three most prosperous regions of the world, viz North America, Western Europe and Japan, within these regions, there is some suggestion for these facilities to be confined to a limited number of industrial districts.

In the case of investment to acquire or monitor the results of new research, this, again, has tended to lead to networks of innovatory activities—which are largely undertaken by MNEs—in the leading industrial countries. Examples include the concentration of high-tech industries along Route 1 in New Jersey and in California's Silicon Valley in the US as well as along the M4 in Berkshire and around the government research laboratories in the UK [20], and in the Baden Württemberg industrial district of West Germany. In addition, new science parks and R&D centers are being set up in the Mayagi and Ibaragi prefectures in Japan [24] and at Trieste, Bari and Biella in Italy; and regional centers for technology diffusion in Emilia and Lombardy, also in Italy [29]. Depending upon the nature of the activity, this has sometimes led to a dispersion of R&D (e.g. of Swedish ball-bearing companies in Western Europe) or to a concentration of activity (e.g. of US aerospace research in California). The number of R&D-related foreign activities by MNEs of all nationalities has, however, increased.

2.6 THE IMPACT OF MNEs ON INNOVATORY CAPACITY OF HOST COUNTRIES

The above paragraphs have suggested that MNEs have particular advantages over non-multinationals in R&D, just as they do in some other value adding activities. These advantages stem from their size, accumulation of technological know-how and geographical spread, which enable them both to better diversify risks and to coordinate the inputs and outputs of R&D more effectively. Moreover, the setting up of branch R&D laboratories will not normally have to incur the same setting up of costs of a *de novo* R&D unit, as would a local producer.

The evidence suggests that recent changes in organizational and locational variables have had an ambiguous affect upon the amount and kind of foreign R&D activity by MNEs. As a whole, it would seem that there has been a modest increase in the proportion of foreign R&D expenditure undertaken by US and European multinationals over the past decade, particularly in developed countries. The research intensity of foreign-based production has also risen. Between 1977 and 1989, for example, the R&D expenditures by US foreign affiliates rose from 0.41% to 0.79% of sales; while those by foreign affiliates in the US rose from 0.48% to 1.17% (in 1988) [43, 44]. Employment in R&D activities by US firms abroad increased from 63.0 thousand (or 1.2%) of total employment in 1977 to 95.2 thousand (2.0%) of total employment in 1989. Finally, it would seem that, in the case of advanced industrialized nations, at least, both the share of R&D undertaken within these countries, and that by their own MNEs in foreign countries has risen. Between 1977 and 1982, for example, industrial R&D expenditure in the five leading industrialized countries[25] rose on average by 189.2% (in money terms), whereas the R&D expenditure by foreign affiliates in those countries increased on average by 286.1%. In both Japan and the US, the share of industrial research undertaken by foreign affiliates[26] more than doubled over that period.

Data on the share of US registered patents attributable to the foreign affiliates of UK, French and German MNEs, as a percent of the total patents registered by all firms in these countries, tell a similar story. In the period 1983

---

[25] Viz the US, UK, West Germany, France, and Japan.
[26] In Japan's case, US affiliates. Data are derived from Table A61 of National Science Foundation [31], the 1977 and 1982 Benchmark surveys on the foreign activities of US firms ([43]), and the 1980 and 1982 Surveys on foreign direct investment in the US [44]).

MNEs and the Globalization of Innovatory Capacity ——————— 35

to 1986, 15.2% of patents attributable to research in the UK were owned by foreign affiliates, compared to 10.9% in the period 1969 to 1972; the corresponding contributions of foreign affiliates in France were 8.9% and 6.3%; and of those in West Germany 8.8% and 5.6%. These figures are broadly consistent with some estimates by Stoneman [40] of the proportion of R&D expenditure by business enterprises in the UK, France and Italy in 1985 accounted for by non-resident firms. The respective percentages were 10.9%, 7.4% and 6.5%.

We have observed that Japanese firms have only recently begun to engage in R&D activities abroad, although, like many third world MNEs, they have established research oriented listening posts, particularly in the US [21]. However, there are some indications that, because of their perceived need to have a research presence in Europe and North America, they will internationalize their R&D facilities sooner, and perhaps more extensively, than their American and European counterparts[27].

We now turn to consider some possible effects of globalization of MNE activity on the innovatory capacity of home and host countries. We shall divide our discussion into the direct and indirect effects.

2.6.1 Direct Effects

The direct effects will be dealt with briefly. They may be identified as the amount of innovatory capability actually created by the affiliates of MNEs in a host country, less the amount which would have been created in their absence. The first component we have considered in the previous section. The second depends entirely upon the assumed opportunity cost of inward direct investment, and, in particular, the extent to which foreign affiliates are presumed to replace innovating capacity by indigenous firms or supplement such capacity. The former might (but will not necessarily) occur when a MNE acquires or out-competes a domestic (or potential) domestic competitor. Where the competitor is, or would have been, undertaking more R&D than the affiliate, e.g. where a subsidiary only undertakes Type 1 R&D, while a local firm would have undertaken Type 2 R&D, then the *net* direct effect on technological capacity may be negative. In addition, or alternatively, the kind of innovatory activity performed by the two groups of firms may be different. For example, a foreign affiliate might undertake specialized (Type 3) research, and an indigenous competitor might undertake basic

---

[27] According to JETRO (quoted by Herbert [21]) there are more than 100 Japanese affiliates in the US which engage in R&D activities. Most of them tend to be clustered in California or in the Boston–Washington industrial corridor. At the end of 1990, only six Japanese subsidiaries in the UK had their own R&D facilities. These were Nissan European Technology Centre, Canon Research Centre, Kobi Steel, Sharp Laboratories of Europe, Hitachi, Yamanouchi Pharmaceuticals and Sony.

(Type 2) research. Only in the case of a greenfield research venture engaging in Type 4 R&D would it seem, that in the absence of foreign investment, there would be less direct research in the host country. However, as we shall see, the secondary effects of MNE activity may be very different.

Empirical evidence on the innovating *diversion* and *creating* effects of inward direct investment is extremely limited; indeed, for all intents and purposes, it is non-existent. Moreover, its short and medium term consequences may be different from its long term impact. Insofar as most greenfield manufacturing investment initially leads to a net increase in value added activities of the host country, it may be reasonable to hypothesize that at least some of these might eventually lead to an increase in domestic innovatory capacity. But it is entirely possible that, by the time a subsidiary begins to undertake R&D, it is cutting into the markets of its competitors and, hence their ability, or incentive to engage in R&D.

One useful measure of the impact is the actual amount of R&D undertaken in a particular industry, where MNEs are present and where they are not. Again, in practice, such controlled experiments are extremely difficult to conduct. However, it is possible to observe changes in the innovatory capacity of an industry over time, and the apparent impact of inward foreign investment on that capacity. This, the following section tries to do. For the moment, we would observe that such an impact may arise as much from the *control* exerted by a MNE over the deployment of the human and technological capabilities it transfers, as from the capabilities themselves. This is because this control may affect not only the innovatory competencies of the affiliate, but that of local producers and consumers as well.

So far in our analysis, we have been concerned with the effects of inward direct investment. But, equally, it may be that home country multinationals undertake R&D in foreign countries which might otherwise have been undertaken in their home countries. If R&D is undertaken by UK multinationals in the US, does this mean that R&D capacity in the UK is less than it would otherwise have been? Again, the answer depends upon the kind of R&D and the opportunity cost of such investment. It also depends upon the economic and political environment in which firms operate. Firms respond to market signals and to government actions. If these cause them to initially locate R&D in a foreign country, then it may be that a change in the value of these variables would induce them to behave differently. While MNEs may fashion external events, they are also influenced by them. Since the location of, at least, some R&D activities by MNEs are determined by country-specific characteristics, e.g. the availability and cost of scientists and engineers, which may be government controlled or influenced, it follows that government policies may be a critical factor affecting both ownership and the location of R&D. We shall take up this point further in Section 2.8.

2.6.2 Indirect Effects

Perhaps, however, the main impact of MNEs on the globalization of innovatory capacity rests upon the secondary or spill-over effects of these activities. Such effects are of three main kinds. The first is on the suppliers and customers of the foreign affiliates; the second is on their indigenous competitors; and the third is on other employers via the mobility of labor or through normal commercial intercourse. As to the first effect, the relevant question is the extent to which foreign affiliates either create a demand for intermediate products which require new or up-graded innovatory capabilities, and/or supply intermediate products which, if they are to be used effectively, need to be used jointly with other new or up-graded resources. For example, a foreign-owned motor vehicle assembler may create the demand for higher quality or new types of components, or negotiate a lower price for existing components. This may lead either to the creation of new technological capacity by the supplier, or force it to make more efficient use of its existing capacity.

The literature is replete with examples of the backward linkage effects of multinational affiliates [42, 27]. In some cases, these can be very positive indeed. New information, financial assistance, technological advice, quality control and inspection methods have all been shown to be important in a variety of industries, e.g. in the motor vehicle and consumer electronic sectors. On the other hand, foreign affiliates may reduce indigenous technological capacity if they import a higher proportion of intermediate products than their local competitors. Indeed, there are instances, where, in order to stay competitive, domestic component suppliers have themselves migrated overseas [10]. In this case, a 'vicious' technological circle may be set in motion by inward direct investment, the end result of which is that technological capacity—at least in the sector to which the investment is directed—is eroded[28].

Much, of course, is likely to depend upon the kind of inward investment undertaken by MNEs and the inducements offered by host governments for such companies in their midst. It is now becoming increasingly accepted that the stronger the innovatory infrastructure of a country, the greater the rivalry between firms, the greater the pressure exerted by consumers on producers to innovate new products, and the more pronounced the indigenous entrepreneurial ethos is, then the more likely it is that foreign owned firms will wish to engage in innovatory activities[29]. At the same time, as the

---

[28] The concept of vicious and virtuous technological circles is explored at length in [5]. However, it should be observed that, from a macro-resource allocative viewpoint, it may be entirely appropriate that the host country *should* restructure its innovatory activities.
[29] In the study of the globalization of R&D undertaken at the University of Reading, about one-half of the 135 affiliates providing information indicated that they provided technical support for their suppliers. See [8].

previous section has shown, firm- and industry- specific characteristics are also important in influencing the kinds of R&D undertaken by MNEs outside their national boundaries.

The same logic might be used to explain the impact of foreign affiliates on their industrial customers. Where such affiliates either enable intermediate products to be supplied more efficiently than otherwise would be the case, or, by one means or another, assist their downstream customers to use these intermediate products more effectively, then it is likely that their technological capabilities will be advanced. Examples include MNEs in agribusiness and petroleum refining sectors, which have helped downstream customers to engage in high value secondary processing activities; and those which produce motor vehicle and durable consumer goods, where there may be an important element of after-sales servicing involved.

The effect on competitors will also be varied. In some cases, inward investment may be powerful enough to drive out indigenous competitors, and with it their R&D activities. Where this occurs, but the investing company continues to concentrate its innovatory expenditures in its home country, the host country will lose. In other cases, the foreign firms will stimulate their indigenous rivals to become more innovatory or to better deploy their existing R&D resources. This may have two consequences. The first is that there will be a redistribution of R&D between foreign and locally owned firms. The second is that the competitiveness of the domestic economy is likely to be improved *vis-à-vis* that of other countries. If the latter event occurs, inward investment may have been the driving force, via its impact on indigenous innovatory capacity. However, it is worth observing that manufacturing and innovatory activities are not necessarily spatially linked. It could be, for example, that a US or Japanese MNE chooses to undertake most of its European manufacturing activities in one country and its R&D activities in another.

A similar type of analysis can be applied to outward R&D activity by MNEs. In some situations, particularly where the R&D is intended to take advantage of superior innovating facilities of the host country, this may help to strengthen the global competitive position of the investing firms and, hence, make it more likely that, in the long run, their domestic technological capabilities are improved. In other cases, where the setting up or expansion of foreign R&D activities is driven by structural market distortions or by unfavorable supply capabilities in the home country, then that country—depending upon what happens to the resources released—may be worse off. Investment designed to gain access to, or an insight of, foreign R&D activities may, in the long run, be beneficial to home country innovatory capacity; just as inward investment, for the same purpose, may take on the characteristics of a Trojan horse.

In conclusion, the indirect effects of foreign-based R&D by MNEs may be widespread and significant, but they are difficult to generalize about. Much depends upon the market conditions in which the R&D is undertaken and the eco-structure of both home and host countries. There is support both for the proposition that inward foreign investment may lessen indigenous innovatory capacity and for the proposition that it will increase it. Similarly, cases to support the hypothesis that outward investment in R&D is *complementary* to domestic R&D can be found alongside those which suggest that the two kinds of R&D *substitute* for each other. What does seem very plausible is that where foreign production adds to domestic production, the R&D base of the investing company is strengthened—whatever the nationality of the firm.

2.7 TWO CASE STUDIES

Because of the paucity of statistical data, it is not possible to come to any definitive conclusions as to the effect of multinationals on indigenous technological capacity. But, industry case studies do offer some useful pointers. Some years ago, the OECD examined this very question in respect to the pharmaceutical, food and drink, and computer sectors in Western Europe [32, 33, 34]. It found that the answer varied according to both industry- and country-specific factors. In the pharmaceutical industry, outward and inward investment were found to have had a positive impact upon the innovatory capacity of countries wherever domestic rivalry and technological infrastructures were strong, where outward investment was based on the competitive strength of the investing firms rather than on the locational weaknesses of the home countries, and where inward investment was based as much on the innovating capabilities of the host countries as on the competitive advantages of the investing firms. Countries such as the UK, France, Germany, and Switzerland possessed these characteristics. Countries which had only a limited innovatory capacity or offered weak local domestic rivalry to inward investment usually had their technological base eroded as a result of inward investment. Examples included Norway, Denmark, and Portugal.

Similarly, outward investment was shown to aid indigenous innovatory capacity where it led to additional output (and hence strengthened the sales base for domestic R&D), helped firms acquire (at a competitive price) R&D facilities, or enabled domestic firms to better monitor the innovatory activities of their foreign competitors, where the investing firm takes advantage of its geographical diversification to promote an R&D specialization.

The British *motor vehicle* and *pharmaceuticals* industries offer interesting examples of the very different interactions which may occur between inward investment and indigenous firms. For most of the two decades up to 1987, the UK was not an attractive European location for motor vehicle production; indeed the industry had one of the poorest records of productivity in the industrialized world. Consider a couple of very telling statistics. The UK content of cars sold in the UK by the multinational producers fell from 88% in 1973 to 46% in 1984 in the case of Ford, 92% to 42% in the case of Talbot and 89% to 22% in the case of General Motors. Over the same period, the motor vehicle industry experienced the greatest deterioration in its trading capabilities of any British manufacturing sector. Between 1978 and 1984, for example, vehicle exports rose by 13% (in value terms) while imports increased by 118%. In volume terms, the exports of all the major UK producers fell quite dramatically in the early 1980s. Britain's share of the global research and development expenditure by motor vehicle firms, which had always been lower than her share of global production or consumption of motor vehicles, slumped even further after the mid-1960s. Until the mid-1980s, at least, the motor industry was not one of Britain's success stories.

Moreover, as inward investment fell, and became less directed to high value activities, this further weakened the industry's ability to compete. Foreign (and particularly US) MNEs were choosing not to produce in Britain because the supply capabilities and related support facilities were perceived as being inadequate, and because of the fragility of the eco-structure, particularly as revealed by extremely poor industrial relations. Only the most generous incentives and the most persuasive efforts of the UK government induced companies like Ford to make the huge new investments they did in Wales and North West England.

However, there is an interesting sequel to this story. In 1986, Nissan announced that it was going to set up a motor vehicle assembly plant in Washington, Tyne and Wear. This was the first positive sign of investment in an industry which had lost ground to its foreign competitors in the preceding 20 years, and would probably not have occurred had there not been a change in the political and economic climate in the UK; and, more especially, a renewal of more market oriented economic policies. The early years of the Nissan venture have been very successful—so much so that analysts are now talking about a resurgence in the British motor vehicle industry.

In particular, such analysts point to three events. First there has been an acceleration of the assembling program of Nissan; in 1990 nearly 200 000 cars were produced in the UK, compared with 150 000 originally intended by that date. Second there has been an increase in the local (EC) content of Nissan cars—from 50% in 1986 to 70% in 1989, and this is likely to rise to 80% by the end of 1991. In part this has been made possible by new investments by Japanese and German component suppliers in the UK (often jointly with the UK firms), and by Nissan working with UK-owned suppliers to help them improve their product quality and their technological capacity[30]. Third, and perhaps most important of all, the Toyota company has announced that it intends to build substantial assembly and engine plants in the UK. If this comes to pass, the effects could be even more far reaching than those arising from the Nissan venture, as Toyota has a more extensive network of suppliers than any other motor vehicle assembly in the world. Coupled with the BL–Honda venture, inward investment (this time from Japanese and German MNEs) would appear to have stemmed the tide of the 'vicious' circle. Statistics do not yet reflect the change, but an industry which once looked as if it was dying now appears to have been given a new lease on life[31].

As yet, however, the same cannot be said of UK outward investment in the UK motor industry, with the possible exception of Jaguar; although some business analysts claim that, without its Continental European production, the UK motor vehicle component industry would not be alive today. In other words, given the perceived locational disadvantages of the UK at the time, the domestic innovatory base of the components industry was better served by the relatively more efficient Continental production facilities.

By contrast, the tale of the pharmaceutical industry is one of almost uninterrupted vitality and vigor. From the early 1950s, foreign-owned producers in the UK have acted as a competitive challenge to an already quite strong and diverse indigenous sector. The reputation of the UK scientific community, its professional system of drug registration, its patent system, and its clinical testing procedures are all excellent. Its industrial relations and contacts with universities and the medical profession are second to none; and, until recently at least, the government, through the PPBS scheme, has offered a stable and fair reward system for the pharmaceutical companies, as well as providing them with adequate incentives to engage in innovatory activities[32].

The results are shown in a variety of indices. First, the pharmaceutical industry is a substantial net contributor to the UK balance of payments; in 1987 it exported nearly twice the amount it imported. Second, the UK is one of the two or three leading centers for pharmaceutical research and development in the world. At the end of the 1970s, it was estimated that, while only 3.5% of the world's drugs were bought by UK consumers, 5% of the

---

[30] In the US, where Japanese-owned affiliates now account for 10% of the American production of cars, there are more than 300 motor car component suppliers in which Japanese firms have direct investment.
[31] Professor G. Rhys of Cardiff, a leading authority on the international automobile industry, predicted that by the mid-1990s, and largely due to the presence of Japanese auto companies, the UK will once again be exporting as many cars (in value terms) as it imports.
[32] The favorable environment for pharmaceutical R&D in the UK is in contrast with that in Japan, where the pharmaceutical industry has been faced with unfavorable price regulations, imposed by the health care delivery system, and little government support for R&D [38].

world's production was undertaken in the UK, as was 11.5% of the industry's global research and development. Success breeds success. Though foreign companies now account for about 45% of the drugs supplied by the National Health Service, this figure has remained fairly constant over the years. British MNEs are among the world's leading drug producers and in the forefront of innovatory advances. In 1969–77 the revealed patenting advantage of UK based pharmaceutical firms was 0.96; in 1978–86 it was 1.72. Between them, foreign and domestically owned firms have generated a critical mass of physical and human innovatory capacity, which separately neither could have achieved. The agglomerative economies so gained, and the domestic rivalry so generated, are in total contrast to the events which occurred in the motor vehicles industry.

As with the vehicles industry, there is a footnote—and in this case a less welcome footnote—to the competitive position of the UK pharmaceutical industry. In recent years, there has been a cutback in government funding of R&D activities—and particularly those which are university based—in the pharmaceutical and related sectors. In addition, new restrictions have been placed on the kind of drugs which doctors are able to prescribe to National Health Service patients, while the government (as the main purchaser of ethical drugs) has lowered the prices it is prepared to pay for certain products. This has caused some degree of concern, both by foreign and domestic pharmaceutical firms, as to the likely future economic and scientific ambience for R&D in the UK; and a reconsideration by them of the location of new innovatory activities, particularly by US companies, in the EC[33].

We have illustrated from just two UK industries, and no doubt there are equally good examples of 'vicious' and 'virtuous' technological circles in other industries and in different sectors in other countries[34]. We conclude this section by summarizing five conditions in which outward and inward direct investment may strengthen the innovatory base of the home or host country. The first is the availability, or potential availability, of indigenous resources for innovation. The second is a congenial economic environment and ethos to savings, entrepreneurship and risk taking. The third is a strong impetus to upgrading products and productivity provided, on the one hand, by strong inter-firm rivalry and, on the other, by consumers demanding new and better quality goods and services. The fourth is the availability of clusters of vertically or horizontally related economic activities, whose combined value added is likely to be greater when they are undertaken in the

---

[33] From information collected in a project undertaken by the Economist Advisory Group in 1988 into the future of the pharmaceutical industry in Western Europe.

[34] Some examples of the innovatory strengths and weaknesses of different industries in the US, UK, Japan, Korea, Italy, and Germany are given in [37].

MNEs and the Globalization of Innovatory Capacity 43

same country, or a region within a country, than when they are undertaken apart[35]. The fifth is the role of government policy in shaping these forces (which we shall discuss in Section 2.8).

It is the interface between the activities of MNEs and these elements of the eco-structures in which they operate which will determine their impact on the innovatory activity of home and host countries. This impact is also likely to be country-, industry-, and firm-specific. The effects of US MNEs on the capability of the Portuguese economy to generate new kinds of information technology may be very different than those of Japanese investment on the innovatory capabilities of the US motor vehicle industry, or of those of Dutch investment in the Singaporian electronics industry on the upgrading of human and physical resources.

Similarly, the consequences for the innovating capacity of the home country of setting up R&D facilities by US computer firms in Japan, induced by tariff or non-tariff barriers, may be very different from those of the acquisition of Indian or Nigerian R&D laboratories in food processing by UK investors; or those of the setting up of a scanning and monitoring R&D engineering facility in the US by a Korean MNE on Korean indigenous capabilities.

Taken together, then, and in the right circumstances, FDI can enhance and improve the quality of a nation's technological and scientific capabilities. In the wrong circumstances, it can weaken them. The trick, then, is to identify both the types of multinational activity and the conditions in the home and/or host country necessary to bring the former, rather than the latter, situation about.

2.8 THE ROLE OF GOVERNMENT POLICY

Of all the exogenous factors influencing the impact of MNEs on the creation and location of innovatory capacity, and of the dissemination of its output, none, perhaps is more important than the attitudes and actions of the governments of the countries in which they operate. It is well known that, by regulation, commissioning, or funding R&D activities, both national governments and regional authorities may *directly* influence the level and structure of innovatory-related activities[36]. Somewhat less appreciated is the

---

[35] For some examples of vertical (as opposed to horizontal) virtuous and vicious value added circles, see [29].
[36] For example, in 1986, the percentage of national R&D spending funded by government varied between 12.6% in Switzerland and 21.2% in Japan to 56.3% in Italy (in 1987) and 50.8% in the UK. The percentage of government R&D spending on defense varied from 0.5% in Denmark (in 1985) and 3.5% in Japan to 68.1% in the US (in 1988) and 49.2% in the UK [35]. For an examination of the role of the European Economic Community in funding and commissioning EC based technology collaboration, see an excellent Harvard case study by Gomes Casseres and Levy [17]. See also the various publications of the Commission of the European Communities, DGXII.

indirect role which governments can play in fashioning the eco-structural dimensions within which firms are induced to undertake such activities; and in some cases, *which* firms undertake these activities.

Numerous studies have identified the multifaceted role of government in influencing the supply of trained manpower, through educational and vocational training programs; the availability of finance capital, through interest role and fiscal policies (especially towards savings levels); the provision of transportation and communications infrastructure; the degree of rivalry between innovating, or potentially innovating, firms; the market structures within which firms operate, including the ease or difficulty with which they may conclude cooperative alliances; purchasing standards; entrepreneurial incentives and the work ethos, and so on. Each of these policies, in its own way, may influence the ability and motivation of both domestic firms and foreign subsidiaries to upgrade their value adding activities. However, of no less importance is the way in which these policies are dispensed and integrated with each other. Let us explain what we mean.

For the most part, in advanced industrial societies at least, not only are government policies geared towards achieving a different mix of economic and social objectives, but rarely, in the past, has the upgrading of national resources and capabilities, and the promotion of competitiveness been a key objective. In other words, although it may be recognized that competition, education, fiscal, and environmental, and other policies may affect the innovatory capabilities of the country, this has not generally been the main focus of such policies. Viewed from the particular perspective of advancing competitiveness, such policies have often appeared piece-meal, uncoordinated, and ineffectual. Moreover, the policies of governments to *directly* influence innovatory capacity may often be less successful than they might be simply because the rest of government policy is not conducive to their success.

It is not the purpose of this chapter to make a comparison between the attitudes and policies of different governments, which may, directly or indirectly, affect innovating capacity and the role played by MNEs. But, of the diversity of such attitudes and policies, and of their effectiveness, there can be little doubt. The contrast between the holistic and coordinated economic strategies of the Japanese and the Koreans, geared towards a systematic upgrading of their resources and innovatory capabilities, and the fragmented and uncoordinated micro-economic policies of more Western nations, has been well documented[37]. Even the organization of R&D policies may, itself, differ. Henry Ergas [15], for example, distinguishes between the 'mission' oriented technological policies of the US, UK, and France and the 'diffusion' oriented policies of Germany, Switzerland, and Sweden; which he

---

[37] For a very recent account see [37].

MNEs and the Globalization of Innovatory Capacity 45

suggests will have very different effects on the types and structure of innovatory capacity.

Much less attention has been paid to policies towards the creation or diffusion of innovatory capabilities by foreign firms; or indeed to the location of R&D by domestic MNEs. Yet we have seen that, for the leading innovatory countries, at least, both the proportion of US patents registered, which are accounted for by foreign-owned firms, and the proportion of the patents registered by their own firms attributable to research outside their home countries, are rising [7].

Why should it matter who owns or controls the innovatory capacity of a country? Consider first the case of inward investment. The simple *economic* answer is that as long as the social rate of return from inward investment (which is equal to the value added less profits accruing to the foreign owners) is greater than the opportunity cost of the resources used, it is likely to be beneficial. This condition is easier to identify than to measure, particularly over any length of time. For the critical issue is not so much whether or not the activities of inward direct investors yield a higher domestic value added than that of indigenous firms; or whether that of outward investment earns a higher rate of return than that of domestic investment; but of how, through their product and innovatory policies, contacts with suppliers and customers, competitive stimuli and entrepreneurial example, foreign and domestic MNEs may better upgrade indigenous capabilities and advance economic welfare compared with domestic or uninational firms.

The basic concern over the impact of inward investment on innovatory capacity may be illustrated with reference to the current wave of Japanese manufacturing investment in the UK. Here there appear to be two main anxieties. The first is lest the Japanese undertake only low value added activities in the UK, and centralize their innovatory activities in Japan. Assuming the Japanese affiliates to be more competitive than their indigenous competitors, the innovatory base of the UK then becomes eroded, which, so the argument goes, helps the Japanese to further strengthen their indigenous capacity and Japanese MNEs become even more competitive. One country's vicious circle then becomes another's virtuous circle.

The second anxiety has less to do with the way in which Japanese MNEs may or may not control the amount and kind of resources transferred, and more with the way in which these are used. The argument is that what is perceived good for the global objectives of Japanese multinationals is not necessarily good for the development of the UK economy.

Now, these concerns are not new; indeed, as a student of international investment since the 1950s, I have a strong sense of *déjà vu*! But, because of the increasing importance attributed by countries (particularly advanced industrialized countries) to continually upgrading their resources to improve their competitiveness, the contribution of MNEs to innovatory capacity is being more and more judged from this viewpoint [23].

While these worries are understandable, they are based on a whole set of assumptions which may or may not be true. For example, it is virtually certain that, had not the greater part of Japanese manufacturing investment come to the UK, it would have been directed elsewhere in the EC, and the UK would have been faced with the same competitive pressure from Japanese firms. Secondly, it cannot be assumed that the UK resources used by Japanese investors would have been put to better use elsewhere in the economy, for example by competitive firms. Much depends, in this instance, upon the macro-economic policies pursued by the UK government and the relative efficiency of resource usage by UK and Japanese firms.

Third, we have seen that the presence of foreign-owned firms may not only stimulate indigenous firms to be more efficient, but may help create or sustain centers or clusters of activity which yield agglomerative economies of benefit to competitors. Fourth, even if it could be demonstrated that Japanese investment weakens the innovatory capacity of the sectors in which it invests, it might well be that the resources released would be of even greater value elsewhere in the host economy, and that inward investment performs a useful restructuring function.

Much, of course, depends on the assumptions made about the adequacy of the existing eco-structure for innovatory capacity, and the role of government in formulating resource allocative mechanisms which best lead to the improvement of the quality and use of human, physical and financial resources at minimum cost. Such policies do not only relate to innovatory activity *per se*, but those which in any way affect such activity. Innovatory competitiveness in a modern global economy depends, first and foremost, on the implementation of pro-competitive public policies and on the provision of adequate supportive infrastructures [11].

One of the difficulties in evaluating the economic impact of MNEs on technological capacity, where intermediate product markets are imperfect and distorted, is to know how best to assess the true economic worth of the resources used. Almost all the markets identified contain elements of imperfection. Sometimes, governments add to these imperfections, e.g. by imposing import quotas, offering regional subsidies or regulating prices; sometimes they help remove them, e.g. by anti-trust policies and by insurance guarantee schemes. But, when judging the response of policy makers to inward or outward direct investment, it is entirely possible that the effects are not as beneficial as they might be, either because of market distortions or because of the failure of markets to correctly signal the need for social beneficial innovatory-related activities. Rather than control the activities of MNEs, it may then be preferable for governments to enact policies which, taking cognizance of the globalization of markets, will best enable them to upgrade and restructure their resources to meet the needs of both domestic and international consumers.

We accept, of course, that, however much governments may seek to put their internal economic affairs in order, they may still be faced with distortion in international markets caused by other governments, which act to the disadvantage of their own firms. Where these cannot be stopped or neutralized by international action or negotiation, it is entirely understandable that the disadvantaged governments may wish to respond by some kind of retaliatory action, which may include policies which directly affect the innovatory capacity of domestic and foreign-owned firms.

Thus, if Japanese policy makers do not do their best to free Japanese domestic markets to allow a level playing field between domestic and foreign firms, then, given that market distortion, it may well be in the best interests of European and US governments to negate that particular 'strategic' advantage by imposing (or not getting rid of) discriminatory policies towards Japanese goods or investors producing in their territories.

2.9 CONCLUSIONS

It is difficult to generalize on the direct or indirect effects of MNEs on the innovatory capacity of home and host countries. So much depends on the reasons for the ownership and cross-border location of value added activities of MNEs, and the response of indigenous firms to them. This, in turn, will depend upon the economic and institutional environment in which R&D is organized and located, and the role of governments in shaping this environment and facilitating the upgrading of their indigenous resources.

In the past, countries have adopted different policies towards inward direct investment according to how they have perceived such investment might impact on national economic (and other) goals. Broadly speaking, there are two main views on this impact. The first is that FDI speeds up the process of economic growth and restructuring. It does so both by providing technology, entrepreneurship and organizational skills at a lower cost than any alternative usage of resources, and by its competitive stimulus and spill-over effects on the rest of the economy. The alternative view is that, while in the short run this may, or may not, be the case, in the long run it is only likely to happen if MNEs do not distort (or add to the distortion of) markets; and as long as the control exerted over their affiliates' activities is consistent with the innovatory goals of the countries in which they operate. To a certain extent, the globalization of markets and the growth of intra-industry trade and investment is helping to bridge these views by directing more attention to the actions of governments in setting the conditions in which MNEs are able to efficiently perform the role required of them.

The perception that, because they are faced with market distortions and failures, MNEs will not, left to themselves, ensure the best distribution of innovatory capacity, has dominated the thinking of countries like Korea and Japan. In these cases, not only has government policy towards outward investment been geared to sustaining the domestic innovatory capacity of the home country, but it has deliberately limited inward investment until indigenous technological capability is sufficiently strong for such investment to interact with it in a mutually beneficial way. Germany and Singapore have followed a different strategy. They appear to believe that inward investment is the quickest (and often the cheapest) way of upgrading local technological capability, and that the opportunity cost of such investment is more than covered by the increases in social productivity it brings about. The debate is by no means resolved; and it is being increasingly complicated by the fact that the avenues of creating or obtaining innovatory capacity are widening both for firms and for countries. Joint ventures, strategic alliances, subcontracting arrangements and inter-government cooperation are some of these ways, each of which has its own particular costs and benefits for the participating partners and for the countries involved.

We would make one final point. However understandable it may be that countries wish to advance their technological practices, and however much one might judge the success of inward and outward investment in these terms, all countries cannot expect to be technologically capable in all sectors. The principle of comparative advantage is no less applicable in explaining the international allocation of innovatory capacity than it is in explaining the trade in final goods and services. Unless it is prepared to sacrifice economic welfare for other goals, no country can expect to be entirely self-sufficient in its technological capabilities any more than it can expect to be self-sufficient in all goods and services. So the contribution of MNE activity must be judged, not only in the light of its effects on the *generation* of innovatory capacity, but on the *allocation* of that capacity; and this, in turn, on the long run economic interests of the country concerned. Indeed, it is possible to conceive of a multinational doing a country a service by transferring R&D out of that country, whenever this opens the door to more productive R&D—or for that matter non R&D—activities.

But this raises another question. To what extent do countries wish to be locked in the international division of labor fashioned by MNEs? This question has not been addressed in this chapter, not because we do not consider it to be important, but because we have deliberately limited ourselves to some of the positive (as opposed to normative) implications of the globalization of R&D.

2.10 A NOTE ON THE FUTURE

It is likely that the possession of innovatory capacity and the ability to organize that capability will continue to be an important consideration influencing the competition of both firms and countries in the foreseeable future. It also seems likely that the organization and location of such activities will take on a more pluralistic character. One can foresee a greater amount of intra-firm, intra-industry trade in R&D; with many more cross-border alliances and linkages than in the past. In many respects, we believe that the question of whether or not MNEs have a good or bad impact on technological capacity is no longer of primary interest. Instead, the focus of attention is likely to shift to the identification of the appropriate forms of internationalization of business, the right firms, and the right government policies, which will best enable the innovatory capabilities desired by governments to be achieved.

2.11 REFERENCES

[1] J.A. Alic (1990), Cooperation in R&D, *Technovation* **10**,) 319–332
[2] D. Archibugi and M. Pianta (1989), *The Technological Specialization of Advanced Countries*, Report to the Commission of the European Communities, Technical Report of the National Research Council, Rome.
[3] D. Brash (1966), *American Investment in Australian Industry*, Australian University Press, Canberra.
[4] G.C. Cainarca, M.G. Colombo and S. Mariotti (1988), *Cooperative Agreement in the Information and Communication Industrial System*, Milan Politecnico.
[5] J.C. Cantwell (1989), *Technological Innovation and Multinational Corporations*, Basil Blackwell, Oxford.
[6] J.C. Cantwell (1990), Theories of International Production, in Pitelis, C. and Sugden, R. (eds) *The Nature of the Transnational Firm*, Routledge, London.
[7] J.C. Cantwell and C. Hodson (1990), The Internationalization of Technological Activity and British Competitiveness, A Review of Some New Evidence, in Casson M.C. (ed.) *Global Research Strategy and International Competitiveness*, Basil Blackwell, Oxford.
[8] M.C. Casson (ed.) (1990), *Global Research Strategy and International Competitiveness*, Basil Blackwell, Oxford.
[9] F. Chesnais (1988), Technical Cooperation Agreements Between Firms, *STI Review* No. 4, December, OECD, Paris.
[10] K. Cowling (1986), Internationalization of Production and Deindustrialization in Amin, A. and Goodard, J. (eds), *Technological Change, Industrial Restructuring, and Regional Development*, Allen & Unwin, London.
[11] W. Davidson (1988), *Ecostructures and International Competitiveness*, University of Southern California, Mimeo.
[12] J.H. Dunning (1958), *American Investment in British Manufacturing Industry*, Allen & Unwin, London.
[13] J.H. Dunning (1989), The Theory of International Production in Fatemi, K., *International Trade*, Taylor & Francis, New York.

[14] J.H. Dunning and R.D. Pearce (1985), *The World's Largest Industrial Enterprises*, Gower Press, Farnborough.
[15] H. Ergas (1988), *The Importance of Technology Policy*, Mimeo, Paris.
[16] P. Ghemawat, M.E. Porter and R.A. Rawlinson (1986), Patterns of International Coalition Activity in M.E. Porter, (ed.), *Competition in Global Industries*, Harvard University Press, Harvard.
[17] B. Gomes Casseres and D. Levy (1989), *Technology Collaboration in Europe*, Harvard Business School Case No. N9-389-130.
[18] P. Gugler (1991), *Les Alliances Stratégiques Transnationales*, University of Fribourg Press, Fribourg.
[19] L. Håkanson and R. Nobel (1989), *Overseas Research and Development in Swedish Multinationals*, Mimeo, Stockholm.
[20] P. Hall, M. Brecheny, D. McQuaid and D. Hart (1987), *Western Sunrise*, Allen & Unwin, London.
[21] A. Herbert (1989), Japanese R&D Activity in the US, Research. *Technology and Management*, 32, 11–20.
[22] IMD/World Economic Forum (1989), *World Competitiveness Report 1990*, IMD, Geneva.
[23] B.R. Inman and Burton (1990), Technology and Competitiveness: The New Policy Frontier, *Foreign Affairs*, Spring, 116–34.
[24] Japan Update (1990), *Direct Investment in Japan: New Developments*, Winter, 12–15.
[25] K. Kumar (1990), *Multinational Enterprises in India*, Routledge, London.
[26] S. Lall (1987), Multinationals and Technology Development in the Host LDCs in J.H. Dunning and M. Usui, (eds), *Structural Change, Economic Interdependence and World Development*, Vol. 4, *Economic Interdependence*, Macmillan, Basingstoke and London.
[27] J. Landi (1985), *Vertical Corporate Linkages and Technology Diffusion: The Case of Multinational Enterprise in the Nigerian Economy*, PhD Thesis, University of Reading.
[28] J.R. Lincoln (1992), Work Organization in Japan and the United States, in Kogut, B. (ed.) *Country Competitiveness: Technology and the Organizing of Work*, Oxford University Press, Oxford.
[29] F. Malerba (1990), *The Italian System of Innovation*, Mimeo, Universita L. Bocconi, Milano.
[30] National Science Foundation (1989), *International Science and Technology Data Update 1988*, National Science Foundation Washington.
[31] National Science Foundation (1990), *Science and Engineering Indicators, 1989*, National Science Foundation, Washington.
[32] OECD (1981), *The Impact of Multinational Enterprises on National Scientific and Technological Capacity: The Pharmaceutical Industry* (Report prepared by J.H. Dunning, M. Burstall and A. Lake), OECD, Paris.
[33] OECD (1982a), *The Impact of Multinational Enterprises on National Scientific and Technological Capacity: The Computer Industry* (Report prepared by C.A. Michalet), OECD, Paris.
[34] OECD (1982b), *The Impact of Multinational Enterprises on National Scientific and Technological Capacity: The Food and Drink Industry*, OECD, Paris.
[35] OECD (1989), *Main Science and Technology Indicators*, OECD, Paris.
[36] R.D. Pearce (1990), *The Internationalization of Research and Development*, Macmillan, London.
[37] M. Porter (1990), *The Competitive Advantage of Nations*, Free Press, New York.
[38] M.R. Reich (1990), Why the Japanese Don't Export More Pharmaceuticals: Health Care Policy as Industrial Policy, *California Management Review* 32, 124–150.
[39] A.E. Safarian (1966), *Foreign Ownership of Canadian Industry*, McGraw Hill, Toronto.
[40] P. Stoneman (1989), *Overseas Financing for Industrial R&D in the UK*, Paper delivered to British Association (Sector F), September.

[41] Swiss Association of Entrepreneurs (1985), Research and Development in the Swiss Private Sector (no further details), quoted in Borner, Stuckey, Wehrle and Burgener.
[42] UNCTC (1981), *Transnational Corporation Linkages in Developing Countries*, UNCTC 81, A4, New York.
[43] US Department of Commerce (1985 & 1991), *US Direct Investment Abroad: 1982 and 1989 Benchmark Surveys*, DC, Bureau of Economic Analysis, Washington.
[44] US Department of Commerce, Foreign Direct Investment in the US, US Department of Commerce, various dates Washington DC.
[45] UNESCO (1987 & 1990), *Statistical Year Book*, Paris.
[46] D. Van den Bulcke (1985), Belgium in Dunning, J.H. (ed.) *Multinational Enterprises, Economic Structure and International Competitiveness*, John Wiley, Chichester.
[47] E. Westney (1992), Country Patterns in R&D Organization: The United States and Japan, in Kogut, B. (ed.) *Country Competitiveness: Technology and the Organization of Work*, Oxford University Press, Oxford.

Chapter 3

Large Firms in the Production of the World's Technology: an Important Case of Non-Globalisation*

PARI PATEL AND KEITH PAVITT

US patenting by 686 of the world's largest manufacturing firms shows that their share of the world's production of technology is less than their share of R&D activities, and varies greatly amongst sectors. In most cases, the technological activities of these large firms are concentrated in their home country, the characteristics of which influence the volume and trends in their technological activities much more strongly than the international component of these activities. At the same time, these large firms are major elements in the volume and the pattern of sectoral specialisations in their home countries' technological activities.

3.1 INTRODUCTION

In this chapter, we shall use recently developed data, based on US patenting, to evaluate the importance of the technological activities of the world's largest firms in different sectors and countries. There are at least two interrelated reasons for doing this.

---

* This chapter is based on research funded by the Economic and Social Research Council in the Centre for Science, Technology and Energy Policy, at the Science Policy Research Unit. It was initially published in the *Journal of International Business Studies*, 1991. We are grateful to Sandra Wilson and her fellow students for research assistance, to Richard Dickins for computing assistance, and to John Cantwell, Mike Hobday, Margaret Sharp, Nick von Tunzelmann and three anonymous referees for comments on an earlier version.

Technology Management and International Business: Internationalization of R&D and Technology.
Edited by O. Granstrand, L. Håkanson and S. Sjölander. © 1992 John Wiley & Sons Ltd

The first is that technological change is a central feature in economic development, structural change and improvements in efficiency in all countries. Recent studies have shown that technological activities—as measured through R&D and international patenting—are statistically significant determinants of differences in export and productivity performance amongst the major OECD countries [46, 15, 16]. At the same time, major international differences have emerged over the past 25 years in trends—and subsequent levels—of these activities, both in these countries and in the large firms based on them [35, 17]. Briefly stated, the Japanese have had the strongest upward trend, the Anglo-Saxons (UK and USA) and the Dutch the weakest, and the other continental Western Europeans have grown at rates between the two. By the mid-1980s, the countries spending the highest proportion of national resources on business-funded technological activities were Sweden, Switzerland, FR Germany and Japan, followed at some distance by the USA, and then by Belgium, Canada, France, the Netherlands and the UK.

The second is that the debate continues about the degree to which these technological activities are localised in large firms. The heavy concentration of R&D activities in large firms [18] has led some analysts to conclude that they have a dominant position in countries' development of new technology. On the other hand, a number of studies using other measures show that R&D activities considerably underestimate the volume of technological activities in firms that are too small to have functionally specialised R&D departments [1, 24, 41]. A parallel debate is taking place about the extent and the implications of the international concentration of large firms' technological activities, most often when technology has been made a central explanatory variable in the internationalisation of business[1]. In Vernon's early formulation [50] and in subsequent analyses by Dunning [13] and Cantwell [7], home markets are important determinants of large firms' technological advantage, through the nature and extent of inducement mechanisms that stimulate technical change, and of positive externalities that influence the effectiveness of firms' response to these stimuli.

However, in later formulations [51, 14] Vernon and Dunning suggest that large firms are increasingly footloose in their R&D activities, thereby weakening the links between the development of their technology and their home country. Such trends towards techno-globalism are an essential component of currently fashionable predictions of the emergence of The Stateless Corporation [6]. In the context of these debates, we shall try to answer three questions.

First, how important are large firms in the production of the world's technology? Using US patenting statistics, we show in Section 3.3 that the aggre-

---

[1] See, in particular, [50, 51, 5, 7].

Large Firms and Technology: A Case of Non-Globalisation ———— 55

gate level of importance is less than that shown by R&D expenditures, with considerable variations amongst technological sectors.

Second, how important are the technological activities of large firms in those of their home countries and elsewhere? We show in Section 3.4 the considerable variation both in the relative importance of large firms in their home countries' technological activities, and in the degree of internationalisation of their activities. But in most of the countries at the world's technological frontier, the foreign technological activities of large firms are still not the major feature.

Third, how do the volume, trends and sectoral patterns of large firms' technological activities relate to those of their home countries? In Section 3.5 we show that the two are closely correlated, and that country-specific factors dominate over firm-specific factors.

We begin in Section 3.2 with a description of the nature, strengths and weaknesses of the data base.

3.2 THE DATA SET, ITS ADVANTAGES AND LIMITATIONS

3.2.1 The Nature of the Data Set

The data set has been compiled from information, provided by the US Patent Office, on the name of the company, the technical sector, and the country of origin, of each patent granted in the USA from 1969 to 1986. One difficulty with this source is that many patents are granted under the names of subsidiaries and divisions that are different from those of their parent companies, and are therefore listed separately.

Consolidating patenting under the names of parent companies can only be done manually, on the basis of publications like *Who Owns Whom*. We have now extended our earlier consolidations for the UK and FR Germany [36] to cover 686 of the world's largest firms. With the help of the Economics Department at the University of Reading, we have also included in our data set the following information on each firm: country of origin and sales, employment and R&D expenditures for the years 1972, 1977, 1982 and 1984. Not all these three variables are available for all the firms for each of the years.

Table 3.1 lists the top 20 firms patenting in the USA in the period 1981–86, according to our own consolidated classification, and to the original classification by the US Patent Office. It shows that some firms have very similar numbers of patents in both classifications; in particular, US General Electric, Hitachi, IBM, Toshiba, RCA, Canon, Westinghouse, Dow, Nissan and Mobil. However, other firms have considerably more patents in our consolidated classification, and consequently higher rankings: in particular, Bayer,

Table 3.1 Top 20 patenting firms in the USA (1981–86): Patel and Pavitt list versus the US Patent Office List

Company name	Patel & Pavitt	US Patent Office
General Electric Company (US)	4587	4527
Hitachi	3710	3416
Bayer	3352	2304
IBM	3207	3207
Siemens	3151	2480
Toshiba	3094	2855
Philips Corporation	2968	2464
AT&T	2732	1980
RCA	2716	2716
E.I. Du Pont	2401	1971
Hoechst	2270	1327
Canon	2266	2266
Westinghouse	2145	2090
Ciba-Geigy	1992	1709
Allied Corporation	1989	1085
Dow Chemical Company	1961	1816
Nissan	1960	1887
Mobil Oil	1907	1749
Matsushita	1895	1276
United Technologies	1889	1028

Note: Firms ranked by number of patents in the Patel and Pavitt classification.

Siemens, Philips, AT&T, Du Pont, Hoechst, Allied, Matsushita and United Technologies. At the bottom of the sample, firms' annual sales in 1984 were about $900 million, and average employment about 8000.

Table 3.2 shows the numbers of large firms in our data base, according to their home country and to their principal sector of activity. Just under half the firms are US-owned, about one-fifth are Japanese and just under one-third are European. The UK is the largest European contributor, followed by FR Germany and France. In terms of the industrial distribution, firms with their principal activity in mechanical engineering and metal goods account for 21% of the sample, those in chemicals and pharmaceuticals for 16%, and those in electrical, electronic and computing machinery for 12%.

3.2.2 Advantages and Disadvantages

Patent statistics have been used frequently by economists and other analysts as a proxy measure of technological activities[2]. Their general advantages compared to other measures, such as R&D expenditures, are

---

[2] For a more detailed discussion of the uses and abuses of patenting statistics as a measure of technological activities, see [38].

Large Firms and Technology: A Case of Non-Globalisation ———— 57

Table 3.2 The Distribution of the 686 large firms in the sample by principal activity and country

	US	JP	CA	UK	GE	FR	SE	CH	NL	IT	BE	NO	FI	Ot	Total
Chemicals	35	25	–	2	5	5	–	1	2	2	1	1	–	1 (AU)	80
Pharmaceuticals	18	4	–	3	2	–	–	2	–	–	–	–	–	–	29
Mining (coal & oil etc)	29	10	3	5	4	2	–	–	1	1	1	1	1	–	58
Textiles, cloth. & leather	12	5	–	2	1	1	–	–	–	–	–	–	–	–	21
Rubber and plastics	6	3	1	1	1	1	–	–	–	1	–	–	–	–	14
Paper & wood products	21	6	4	1	1	–	4	–	–	–	–	–	2	1 (IE)	40
Food	33	15	2	14	–	4	1	2	1	–	–	–	–	–	72
Drink and tobacco	8	1	4	8	–	–	–	–	1	–	–	–	–	1 (AU)	23
Non-metallic minerals	11	6	1	6	–	2	–	1	–	–	–	1	–	–	28
Metal manufacture	22	13	6	2	13	4	1	1	1	1	2	1	–	1 (AU)	68
Mechanical engineering	37	12	2	9	6	1	4	2	2	–	–	–	2	–	77
Electrical/electronics	31	18	1	4	4	2	3	1	1	1	–	–	–	–	66
Computing machinery	12	2	–	1	1	1	–	–	–	1	–	–	–	–	18
Instruments	10	6	–	–	1	–	–	–	1	–	–	–	–	–	18
Motor vehicles	12	19	–	3	6	3	2	–	–	1	–	–	–	1 (ES)	47
Aircraft	14	–	–	2	1	4	–	–	–	–	–	–	–	–	21
Other transport	3	1	–	1	–	–	–	–	–	–	–	–	1	–	6
Total	314	146	24	64	46	30	15	10	10	8	4	3	7	5	686

Notes:
(1) Country definitions: US = United States JP = Japan CA = Canada UK = United Kingdom GE = FR Germany FR = France SE = Sweden CH = Switzerland NL = Netherlands IT = Italy BE = Belgium NO = Norway FI = Finland Ot = Others: AU = Austria; IE = Ireland; ES = Spain.
(2) There are two companies where the home country is not easily identifiable: Shell, which we regard as Dutch, and Unilever, which we regard as British.

that—with the advent of modern information technology—they are readily available over long time periods; they can be broken down in great statistical detail, according to firm, technical field and geographical location; and they capture technological activities undertaken outside R&D departments, such as design activities in small firms, and production engineering in large firms. Their main general disadvantage is that, like other routine measures of technological activities, they do not measure satisfactorily one of the major fields of technological growth, namely, software.

The advantages and disadvantages specific to our data base are along three dimensions: the nature of the technological activities measured, variations in the propensity to patent, and the interpretation of trends over time.

3.2.2.1 *The nature of the technological activities measured*

Since a patent is granted normally in recognition of technical novelty, our data are better able to capture technology creation than technology diffusion-transfer-imitation. For those who assume that technology is information (i.e. costly to create, but virtually costless to transfer and reproduce), this distinction is a rigid one. However, in the real world of technology that is complex, partially tacit and specific[3], the diffusion-transfer-imitation of technology generally requires technological activities by the imitator, which sometimes result in improvements over the original[4].

Patenting activities do reflect this type of imitation, which is typical of advanced country companies competing close to the world's technological frontier. However, they do not reflect many other types of imitation and related technological activities not involving originality, such as trade in capital goods and know-how, on-the-job training, assimilative R&D and production engineering, and the foreign education of scientists and engineers. These are particularly important forms of imitation for developing countries [44].

3.2.2.2 Variations in the propensity to patent

Patenting is also an imperfect reflection of novel technological activity. Its primary function is to act as a legal barrier against imitation. Three kinds of variation in the propensity to patent the results of technological activities must therefore be borne in mind.

First, there are variations amongst countries, reflecting differences in the costs (e.g. patenting fees) and benefits (e.g. degree of protection, prospective size of market) of patenting. Patenting in the USA is a reliable metric, since screening procedures are homogeneous and rigorous, and success provides relatively strong protection in a large market. Thus, a recent survey of patenting behaviour by multinational firms shows that the USA is the first foreign country in which they normally seek patent protection [4]. For this reason, the international distribution of the sources of US patenting shows statistically highly significant similarities to the international distribution of business enterprise R&D expenditures, both in aggregate and in specific sectors [48, 47, 33][5]. Second, there are variations in the propensity to patent amongst technical fields, reflecting differences in the relative importance of patenting as a protection against imitation, compared to other factors, such as secrecy, know-how and first-comer advantages on learning curves[6]. For this reason, it is advisable to normalise numbers of patents as a proportion of their respective technical fields.

---

[3] For a fuller discussion, see [11].
[4] For an analysis of the conditions under which this is likely to occur, see [49].
[5] US patenting slightly overestimates technological activities performed in the USA, compared to those performed in other countries, since firms have a higher propensity to patent on home than on foreign markets. It also severely underestimates the considerable volume of R & D undertaken in the USSR and other (former?) centrally planned economies, the efficiency of which is very low in innovation and diffusion, when compared to market economies [23].
[6] For systematic evidence on intersectoral variations in the relative importance of these barriers, see [25, 4].

Large Firms and Technology: A Case of Non-Globalisation ———— 59

Third, there are variations amongst firms in the propensity to patent, reflecting *ex ante* uncertainties and differing patenting practices over the wide range of patents with relatively low value[7]. Nonetheless, statistically significant correlations have been found in the USA between inter-firm differences in R&D, and in US patenting [45, 31].

3.2.2.3 Interpretation of time trends

Given lack of time and other resources, our consolidated classification of the 686 firms has been compiled only for one year—1984. Our time-trend analyses of patenting by companies between 1969 and 1986 therefore reflect the firms as constituted in 1984, and none of the changes resulting from purchases or sales of divisions before or since then. Thus, measured changes over time are composed of those of the parts of the firm retained up to 1984, together with those of acquisitions made up to 1984: in other words, what the firm kept and what it bought, up to 1984.

3.3 LARGE FIRMS IN THE PRODUCTION OF THE WORLD'S TECHNOLOGY

Table 3.3 shows, for 33 technical fields and in aggregate, the shares of US patents granted in 1981–86 to the large firms in our sample, to other firms, to government agencies, and to individuals[8].

3.3.1 Aggregate

In aggregate, our set of large firms accounts for just under half the world's technological activities, as measured by US patenting, and for about 60% of that undertaken by firms. This distribution confirms what we found in an earlier study of the UK and FR Germany [36], namely, a lower concentration of technological activities amongst large firms when measured by US patenting than by R&D expenditures. Although strict comparisons at the world level are not possible, national surveys in OECD countries show that typically about 80% of firms' R&D activities are concentrated in firms with 10 000 or more employees. Given that the cut-off level of employment at the lower end of our sample is about 8 000 employees, the proportion of total patenting accounted for by our large firms would have to be more than 80% to reach the same level of concentration as R&D expenditures.

---

[7] On the varying patent practices of firms, see [4]. On the skew distribution of the value of patents, see [32].
[8] Government agencies are granted patents principally in government-funded R&D programmes in defence, aerospace, energy and basic science. Recent studies in Canada and Italy show that, within the category 'Individuals', are a significant proportion of commercially active small firms [2, 28].

Table 3.3 Sources of US patenting in 33 technical sectors: percentage shares in 1981–86

	Large firms		Govt. agen.	Priv. ind.	Other firms
Semiconductors	80.28	(138)	3.94	2.69	13.08
Hydrocarbons, mineral oils etc.	79.45	(158)	0.82	5.77	13.96
Agricultural chemicals	78.98	(92)	0.96	4.29	15.76
Organic chemicals	77.04	(348)	1.73	2.71	18.52
Photography and photocopy	73.40	(147)	0.39	5.84	20.36
Calculators, computers, etc.	69.23	(281)	1.61	7.14	22.03
Inorganic chemicals	67.37	(218)	2.81	5.57	24.24
Bleaching, dyeing and disinfecting	65.20	(125)	1.94	7.75	25.11
Road vehicles and engines	62.45	(179)	0.34	20.49	16.72
Electrical devices and systems	59.62	(327)	3.26	11.38	25.74
Drugs and bio-affecting agents	59.48	(215)	3.35	8.08	29.09
Power plants	58.17	(153)	2.48	20.79	18.56
Telecommunications	57.41	(289)	6.54	13.69	22.36
Image and sound equipment	57.42	(207)	1.80	17.61	23.17
Chemical processes	56.36	(503)	2.36	10.91	30.36
Plastics and rubber products	55.58	(327)	1.56	14.01	28.84
Metallurgical and other mineral proc.	53.30	(372)	1.75	13.94	31.02
Gen. electrical industrial apparatus	50.30	(407)	2.17	15.73	31.80
Food & tobacco (proc. and products)	48.96	(175)	1.61	15.50	33.92
Non-metallic minerals, glass etc.	48.50	(431)	1.24	20.22	30.04
Mining and wells mach. and processes	47.68	(178)	0.89	22.47	28.95
Nuclear reactors and systems	47.45	(38)	6.83	7.60	38.11
Aircraft	43.05	(62)	14.44	23.47	19.04
Instruments and controls	40.93	(491)	3.55	22.06	33.46
Gen. non-electrical industrial equip.	39.86	(433)	0.97	25.33	33.84
Appar. for chemicals, food, glass etc.	39.76	(516)	0.97	21.42	37.85
Metallurgical and metal working equip.	34.99	(379)	0.68	27.18	37.16
Assembling & material handling appar.	29.97	(377)	0.87	28.85	40.30
Other transport equip. (exc. aircraft)	28.46	(197)	1.39	42.01	28.14
Non-electrical specialised machinery	27.63	(481)	0.76	30.39	41.22
Miscellaneous metal products	23.35	(444)	0.67	40.28	35.70
Other n.e.c.	13.49	(241)	5.25	65.71	15.55
Textile, clothing, leather, wood prod.	13.08	(117)	0.71	52.06	34.15
All sectors	49.10	(660)	2.11	19.68	29.10

Notes:
(1) Table is sorted by the share of large firms.
(2) Each row adds up to 100.
(3) The number of large firms active in technical sector is in parenthesis.

3.3.2 Differences amongst Sectors

Table 3.3 also shows major differences amongst sectors in the relative importance of large firms and of the other sources of the world's technological activities. Government agencies are relatively unimportant in aggregate but

Large Firms and Technology: A Case of Non-Globalisation

Table 3.4 Correlation matrix of various measures of concentration of technological activities: 33 sectors, 1981–86

	Lfirms	Govt.	PInd	OthF	CRSale20
Govt.	−0.040				
PInd	−0.909*	−0.008			
OthF	−0.625*	−0.230	0.273		
CRSale20	0.661*	0.266	−0.564*	−0.576*	
HIPPG	0.606*	0.417	−0.524*	−0.573*	0.806*

Notes:
(1) For each sector:
   LFirms = the share of large firms.
   Govt = the share of government agencies.
   PInd = the share of private individuals firms.
   OthF = the share of firms other than the large firms in our sample.
   CRSale20 = the share of top 20 technologically active firms sorted according to sales.
   HIPPG = Herfindahl Index calculated as the sum of squared shares of the firms active in each technical sector aggregated according to their Principal Activity.
(2) *Correlation Coefficient significantly different from zero at the 5% level.

account for more than 5% in nuclear reactors, aircraft and telecommunications—all technologies heavily influenced by military programmes. As in our earlier analyses, large firms are relatively important in chemicals (eight sectors with shares between 56% and 79%), motor vehicles (62%), and electrical and electronic products (five sectors between 57% and 80%), but unimportant in capital goods (seven sectors between 23% and 40%).

Table 3.4 confirms a significant positive correlation across sectors between our large firms' patenting shares, and the shares of the top 20 technically active firms ranked according to sales; and it confirms a significant negative relationship with shares of patenting of 'Private Individuals'. It also shows that the sectoral shares of 'Other Firms', encompassing the very small up to 8000 employees, are more similar to those of private individuals than to those of our large firms.

3.3.3 Possible Explanation of Intersectoral Differences

Recent analysis has shown that intersectoral differences in the concentration of technological activities can be best understood in the context of dynamic interactions between technological opportunities and their appropriability, on the one hand, and the competitive growth of innovative firms, on the other. Briefly stated, higher technological opportunity and appropriability will result in higher concentration [10, 30, 26]. Both R&D intensive sectors (particularly chemical and electronic products) and capital goods sectors have abundant technological opportunities. One of us has shown elsewhere that the low appropriability and concentration in capital goods is positively related to a greater spread of technological activities in capital goods amongst UK firms with different principal sectors of activity [40, 27].

Our data tend to confirm this pattern. Table 3.3 shows relatively low concentration of capital goods technology activities in our large firms, together with a relatively high proportion of these firms producing some capital goods technology, albeit at a relatively low level. This is reflected in the significant and positive correlations shown in Table 3.4 between sectoral levels of concentration of technological activity, on the one hand, and the Herfindahl index of concentration, aggregated according to the sectors of our large firms' principal activity, on the other.

This is because capital goods technology remains largely mechanical. Important mechanical inventions and innovations can still be made without the specialised equipment and range of formal skills required in chemical and electronic technologies [18]. The spatial and design skills of individuals and small groups remain important sources of technology, as do users with experience in operating capital goods. These competences are spread widely across industries and firms, which provide multiple possibilities of entry into promising areas of capital goods technology, thereby reducing the possibilities of appropriation by first-comers. We shall be considering this explanation in greater econometric depth in a future paper.

3.4 LARGE FIRMS IN HOME COUNTRIES' TECHNOLOGICAL ACTIVITIES

The previous section shows considerable variations amongst sectors in the technological activities of the world's largest firms. In this section, we also show variation in their contribution to the world's leading technology-producing countries.

This emerges from Table 3.5, which uses our data on patenting in the USA in the first half of the 1980s to compare the composition of the technological activities of the 11 countries that account for more than 95% of total OECD R&D expenditures funded by business enterprises, and of total US patenting. The first two columns show the shares of total national patenting in the USA granted to the nationally-controlled large firms, and to the foreign-controlled large firms, in our data base, whilst the third column gives the combined share for the other national sources (i.e. government agencies, other firms and individuals). Thus, assuming that US patenting reflects national technological activities, Table 3.5 shows that 8.8% of technological activity in Belgium came from Belgian large firms, 39.7% from non-Belgian large firms, and the remaining 51.5% from other sources in Belgium (firms, government agencies, individuals).

Large Firms and Technology: A Case of Non-Globalisation ———— 63

Table 3.5 Large firms in national technological activities, 1981–86

Country	National sources of patenting in US (3 columns add up to 100%)			Patenting in US by nationally controlled firms from outside home country
	Large firms		Other	
	Nationally controlled	Foreign controlled		(% of national total)
Belgium	8.8	39.7	51.5	14.7
France	36.8	10.0	53.2	3.4
FR Germany	44.8	10.5	44.2	6.9
Italy	24.1	11.6	64.3	2.2
Netherlands	51.9	8.7	39.4	82.0
Sweden	27.5	3.9	68.6	11.3
Switzerland	40.1	6.0	53.9	28.0
UK	32.0	19.1	49.0	16.7
W. Europe	44.1	6.2	49.7	8.1
Canada	11.0	16.9	72.1	8.0
Japan	62.5	1.2	36.3	0.6
USA	42.8	3.1	54.1	3.2

Note: All columns as percentage of total national patenting in US, 1981–6.

The fourth column shows US patenting by nationally-controlled firms from outside their home country, expressed—like the other three columns—as a percentage of total national patenting in the USA. Thus, again by way of illustration, the technological activities of Belgian-controlled large firms undertaken outside Belgium amount to 14.7% of total technological activities inside Belgium, whilst the equivalent proportion is a massive 82% for Dutch-controlled large firms, and a miniscule 0.6% for Japanese-controlled large firms.

By adding up the first two columns, we can see that the relative importance of our large firms varied from around 30% of national technological activities in Canada and Sweden to just over 60% in the Netherlands and Japan, with the remaining seven countries (and Western Europe taken as a whole) in the range from 36% to 54%. By comparing the first and second columns, we see that the relative importance of nationally-controlled and foreign-controlled large firms varied more widely amongst countries: national firms from more than 60% in Japan to less than 10% in Belgium, mirrored by foreign firms from nearly 40% in Belgium to just over 1% in Japan. Simple correlation tests show that neither the relative importance of large firms in total technological activities, nor the mix between national and foreign ones, is significantly related to country size as measured by Gross Domestic Product.

The fourth column of Table 3.5 also shows even greater variation amongst countries in the relative importance of the technological activities of our large firms outside their home countries, from more than 80% of the national total for the Netherlands, to less than 1% for Japan. On the basis of data for 140 large firms[8], it has been shown that the degree of internationalisation of large firms' technological activities is closely correlated to that of production. The same is true for countries in Table 3.5.

However, a comparison of the first and fourth columns of Table 3.5 shows that, in spite of considerable variations amongst the large firms based in different countries, their technological activities remained far from globalised. Only Belgian and Dutch large firms executed more of their technological activities outside their home country than inside. British, Canadian, Swedish and Swiss firms executed between 30% and 42% abroad, whilst firms from the three largest technological countries—FR Germany, Japan and the USA—performed less than 15% outside, as did France and Italy.

Finally, if we compare the second and fourth columns of Table 3.5, we can conclude that, for most countries, the international technological activities of our large firms are not a dominant feature of national technological systems. Only for the Netherlands and Switzerland did the foreign-executed technological activities controlled by large national firms amount to more than 20% of the national total; and only in Belgium and Canada were foreign large firms relatively more important than national ones. In 8 out of the 11 countries in Table 3.5 (and for Western Europe as a whole), less than 20% of national technological activities were foreign-controlled, and at the same time the technological activities executed abroad by nationally-controlled firms amounted to less than 20% of national technological activities. This is a long way from any 'globalisation' of the world's technological activities.

3.5 LARGE FIRM PERFORMANCE AND COUNTRY PERFORMANCE

Sections 3.3 and 3.4 above have shown that the large firms in our sample produce about half the world's frontier technology, but that their relative importance varies considerably amongst sectors and amongst the 11 frontier countries, as does the degree of internationalisation of their technological activities. In this section, we shall probe more systematically into the links between the technological performance (measured in terms of levels and rates of growth of technological activities) of our large firms, and of these 11 countries. There remain a number of unresolved analytical and policy questions about the effects on a country's technology of the presence of large firms, and about the nature and direction of the interactions between such firms and their home countries. We shall use our data to test a number of

Large Firms and Technology: A Case of Non-Globalisation ———— 65

simple relationships that have not been tested before. Given the complexities of the real world, we still shall not be able to give complete and conclusive answers.

3.5.1 Structure, Internationalisation and Country Performance

There is a continuing debate about the effects of the structure of industry —and of related technological activities—on a country's technological performance. Some argue that heavy concentration and the prevalence of large firms reduce competition and technological pluralism, and thereby results in a lower level of aggregate technological activities. Others argue the contrary, that large firms can more easily mobilise the required range of skills, reach critical thresholds, and deal with risk, thereby resulting in a higher level of technological activities.

Contradictory arguments are also put forward about the effects on countries' technological activities of multinational firms. For some, a high proportion from foreign-controlled multinationals is likely to augment national activities; for others, it is either the consequence or the cause of deficiencies in nationally-controlled activities. Similarly, a high proportion of technological activities undertaken by large firms outside their home countries is for some a sign of strength, and for others a sign of weakness.

Table 3.6 shows that differences in countries' technological performance, measured in terms of business-funded R&D as a percentage of GDP in 1983 (RDGDP), are positively and significantly correlated with differences in performance measured in terms of US patenting per capita (PATPC). Neither performance measure is significantly correlated with shares of national large firms in national technological activities (NLFHSH), nor with shares of foreign large firms (FLFSH). Similarly, neither performance measure is correlated with the extent to which national large firms have internationalised their technological activities (NLFASH). However, improvements in national technological performance, measured as real growth of business-funded R&D between 1967 and 1985, are positively correlated with increasing shares of national large firms, at almost the 5% level of significance.

The considerable amount of remaining variance may be explained by another factor, namely, the influence of sectors of national technological specialisation on the shares of national large firms in national technological activities. This cannot be tested statistically, given insufficient degrees of freedom. However, Table 3.5 shows that, although they are very different in aggregate technological performance, Canada, Italy and Sweden have in common both a revealed technological advantage[9] in capital goods [35],

---

[9] For an analytical discussion, see [12]. Such effects can certainly be observed in the UK: multiple reductions of activity in computers and semiconductors; multiple increases in pharmaceuticals and agricultural chemicals; the positive effects of the National Coal Board in coal-mining machinery [34].

Table 3.6 Structural correlates of national technological performance, 1981–86: 11 OECD countries

	RDGDP	PATPC	NLFHSH	FLFSH	ONFSH	NLFASH
PATPC	0.765*					
NLFHSH	0.578	0.431				
FLFSH	-0.435	-0.553	-0.724*			
ONFSH	-0.421	-0.097	-0.756*	-0.096		
NLFASH	-0.061	-0.036	0.250	0.001	-0.362	
NLFTSH	0.254	0.197	0.701*	-0.373	-0.658	-0.865*

Notes:
(1) For each country:
RDGDP = Business-financed Industrial R&D as a percentage of GDP in 1983.
PATPC = Per capita US Patenting, 1981–86.
NLFHSH = Share of National Patenting in USA by National Large Firms: 1981–86.
FLFSH = Share of National Patenting in USA by Foreign Large Firms active in the country: 1981–86.
ONFSH = Share of National Patenting in USA by Other National Firms active in the country: 1981–86.
NLFASH = National Large Firms US patenting from abroad as a percentage of the National Total: 1981–86.
NLFTSH = Total National Large Firms US patenting (home and abroad) as a percentage of the National Total: 1981–86.
(2) *Correlation Coefficient significantly different from zero at the 5% level.

	GRD	GNLFHSH	GFLFSH	GONFSH	GNLFASH
GNLFHSH	0.670				
GFLFSH	-0.393	-0.356			
GONFSH	-0.445	-0.808*	-0.263		
GNLFASH	0.437	0.347	-0.337	-0.146	
GNLFTSH	0.636	0.724*	-0.414	-0.487	0.898*

Notes:
(1) For each country:
GRD = Growth of Industry-financed Industrial R&D, defined as the proportionate change between 1967 and 1985.
GNLFHSH = NLFHSH in 1981–86 minus NLFHSH in 1969–74.
GFLFSH = FLFSH in 1981–86 minus FLFSH in 1969–74.
GONFSH = ONFSH in 1981–86 minus ONFSH in 1969–74.
GNLFASH = NLFASH in 1981–86 minus NLFASH in 1969–74.
GNLFTSH = NLFTSH in 1981–86 minus NLFTSH in 1969–74.
(2) *Correlation Coefficient significantly different from zero at the 5% level.

and relatively unconcentrated technological activities as in capital goods (see Table 3.3). Similarly, Japan and the Netherlands have very different patterns of aggregate performance, but similar relative technological advantages in concentrated sectors: electronics and automobiles (Japan only).

Large Firms and Technology: A Case of Non-Globalisation

3.5.2 Nationally-controlled Firms and Country Performance

We have shown in an earlier paper the strong correlation between the shares of US patents granted to countries, and the shares granted to their national large firms [37]. But this begs the question of causality: do country characteristics determine the behaviour of their national large firms, or vice versa?

We have argued elsewhere that firm behaviour may be strongly influenced by country-wide factors: the degree to which the national financial system properly evaluates intangible, firm-specific assets accumulated through technological activities; the national system of basic research, and education and training of management and the workforce, that influence the quality of major decisions about technology, and of implementation and learning; and the economic climate—and particularly expectations about growth and profits—that influences firms' propensities to invest in technological activities [35]. On the other hand, it can be argued that large firms are not closely coupled to countries; they think and act in terms of world markets, world sources of finance, and world sources of management and worker skills; in typical situations of uncertainty and oligopoly, their discretionary decisions can have major impacts on the rate and direction of countries' technological activities.

Our data can throw some modest empirical light on this debate. We shall typify the competing hypotheses as 'country-dominated' and 'firm-dominated'. In both cases, we would observe a high correlation between country performance and national large firm performance. However, in a country-dominated system, we would also expect a positive and significant correlation between the performance of the two main component parts of national technological activities, namely, the home-based activities of national large firms, and the activities of other national firms. In a firm-dominated system, we would expect instead a high correlation between the performance of the home-based activities of national large firms, and of their foreign activities.

Table 3.7 shows that aggregate national technological performance is country-dominated rather than firm-dominated. Country performance, measured as business funded R&D as a percentage of GDP in 1983 (RDGDP), or as per capita US patenting in 1981–86 (PATPC), is strongly correlated with the performance of national large firms, measured as per capita US patenting (NLFT), but it is even more strongly correlated with the domestic performance of these large firms (NLFH). It is also significantly correlated with the performance of other national firms (ONF), but not with the foreign performance of national large firms (NLFA). In addition, there is no significant correlation between national large firms' domestic performance, and their foreign performance. Table 3.7 also shows the same relations hold even more decisively in performance in growth of technological activities.

Table 3.7 Correlations between the technological performance of countries and nationally controlled firms

	RDGDP	NLFH	FLF	ONF	NLFA	NLFT
NLFH	0.811*					
FLF	0.159	0.114				
ONF	0.665	0.825*	0.292			
NLFA	0.313	0.432	0.348	0.459		
NLFT	0.720*	0.909*	0.241	0.797*	0.769*	
PATPC	0.765*	0.941*	0.280	0.965*	0.482	0.890*

Notes:
(1) For each country:
RDGDP = Industry-financed Industrial R&D as a percentage of GDP in 1983.
NLFH = per capita Home-based US patenting of National Large Firms: 1981–86.
FLF = per capita Home-based US patenting of Foreign Large Firms active in the country: 1981–86.
ONF = per capita Home-based US patenting of Other National Firms active in the country: 1981–86.
NLFA = per capita US patenting of National Large Firms from abroad: 1981–86.
NLFT = per capita US patenting of National Large Firms (home and abroad): 1981–86.
PATPC = per capita aggregate US patenting for the country: 1981–86.
(2) *Correlation Coefficient significantly different from zero (5% level).

	GRD	GNLFH	GFLF	GONF	GNLFA	GNLFT
GNLFH	0.678					
GFLF	0.121	0.439				
GONF	0.468	0.927*	0.521			
GNLFA	0.595	0.175	0.064	0.103		
GNLFT	0.789*	0.957*	0.416	0.870*	0.453	
GPATPC	0.578	0.980*	0.526	0.982*	0.142	0.929*

Notes:
(1) For each country:
GRD = Growth of Industry-financed Industrial R&D, defined as the proportionate change between 1967 and 1985.
GNLFH = NLFH in 1981–86 minus NLFH in 1969–74.
FLF = FLF in 1981–86 minus FLF in 1969–74.
GONF = ONF in 1981–86 minus ONF in 1969–74.
GNLFA = NLFA in 1981–86 minus NLFA in 1969–74.
GNLFT = NLFT in 1981–86 minus NLFT in 1969–74.
GPATPC = PATPC in 1981–86 minus PATPC in 1969–74.
(2) *Correlation Coefficient significantly different from zero (5% level).

3.5.3 Sectoral Performance

Whilst differences in countries' aggregate technological performance are closely correlated with differences in the domestically based performance of large firms, the same may not necessarily hold in specific sectors, especially those where technological activities are concentrated in large, multinational firms. We therefore ran correlations, similar to those in Table 3.7,

Large Firms and Technology: A Case of Non-Globalisation ——— 69

for each of the 33 sectors shown in Table 3.3, with the performance measures being levels and rates of change of per capita US patenting.

What emerges is a pattern remarkably similar to the one in Table 3.7. Only in five chemical and chemical-related sectors and in power plant are there strong correlations between country performance and the domestic performance of nationally-controlled large firms, on the one hand, and their foreign performance, on the other: agricultural chemicals, pharmaceuticals, organic chemicals, dyestuffs and food. Even in the last three of these sectors and in power plant, performance is also correlated with that of other domestic firms.

In 27 out of the 33 sectors, country differences are significantly correlated with differences in firms' domestic activities, but not their foreign activities. These sectors comprise all capital goods, materials, transport, and electrical and electronics. In two sectors—hydrocarbons and motor vehicles—national performance is significantly correlated only with the domestic activities of national large firms; and in one sector—textiles—it is significantly correlated only with other domestic firms. In the other 24 sectors, national performance is significantly correlated with both. In none of the sectors is national performance significantly correlated with that of foreign large firms.

Finally, the domestic performance of large firms is significantly correlated with that of other domestic firms, in about half the sectors, comprising relatively unconcentrated capital goods sectors, materials, and the concentrated sectors of electronics. Linkages between the performance of these two major elements of national technological activities could be of two types. Horizontally, rivalrous behaviour may lead to imitative increases or decreases in technological activities in certain product fields. Vertically, vigorous technological activities in large users of capital goods may induce a complementary response amongst suppliers[10].

3.5.4 Firm Specialisation and Country Specialisation

This leads to the last element in our analysis, namely, the interactions between the sectors of technological specialisation of national large—and other categories of—firms, and those of our 11 countries. These are shown in Table 3.8 which correlates, for each of the 11 countries in Table 3.5, their revealed technology advantage (RTA) in 1981-6 in each of the 33 technological sectors in Table 3.3, with the RTAs of the various categories of firm in Table 3.7[11]. The following conclusions emerge.

---

[10] Revealed technology advantage (RTA) is defined as the share of a country (or firm, or category of firms) in US patenting in a sector, divided by the share of that country (or firm, or category of firms) in US patenting in all sectors. Some readers will note the similarity to the measure of 'revealed comparative advantage' used in analyses of international trade.
[11] As an index of specialisation, revealed technology advantage (RTA) corrects for differences amongst countries, categories of firms, and sectors, in the total volume of patenting activity.

Table 3.8 Sectoral specialisations in technological activity: correlations of RTA indices for 11 countries across 33 sectors, 1981–86

Country	Country and NLT	Country and ONF	Country and FLF	NLFH and ONF	NLFA and NLFH	NLFA and Country
United States	0.88*	0.68*	0.32	0.35	−0.11	0.18
Japan	0.89*	0.68*	0.68*	0.54*	0.85*	0.71*
Canada	0.58*	0.94*	0.54*	0.48*	0.33	0.35
Belgium	0.49*	0.48*	0.72*	0.16	0.59*	0.30
FR Germany	0.90*	0.25	0.26	−0.05	0.57*	0.48*
France	0.44*	0.83	0.06	−0.05	−0.19	−0.30
Italy	0.62*	0.90*	0.29	0.33	0.27	0.15
Netherlands	0.68*	0.29	0.35	−0.16	0.61*	0.51*
Switzerland	0.93*	0.22	−0.08	0.05	0.55*	0.48*
Sweden	0.78*	0.70*	0.02	0.18	0.45*	0.46*
United Kingdom	0.57*	0.27	0.04	0.09	0.44*	0.17

Notes:
(1) See footnote 10 for a definition of the RTA Index.
(2) For each country, the RTA Indices are for:
Country = All US Patenting.
NLFH = Home-based US Patenting of National Large Firms.
FLF = US Patenting of Foreign Large Firms active in the country.
ONF = Home-based US Patenting of Other National Firms active in the country.
NLFA = US Patenting of National Large Firms from abroad.
NLFAT = US Patenting of National Large Firms (home and abroad).
(3) *Correlation Coefficient signigicantly different from zero (5% level).

First, Table 3.8 shows that the sectoral patterns of technological advantage of large firms and their home countries are significantly similar. The sectoral RTAs of all 11 countries are significantly correlated with those of nationally based large firms (NLFT). This is the dominant relationship between firm and country specialisations in FR Germany, Netherlands, Switzerland, Sweden and the UK; whilst the RTAs of other national firms (ONF) are more highly correlated with countries' specialisations in Canada, France and Italy.

Second, the sectoral specialisations of foreign large firms (FLF) are strongly correlated with those of their host countries in Belgium and Canada where, as we saw in Table 3.5, they account for a larger share of national technological activities than national large firms. Otherwise, there are significant correlations between the two in Japan.

Third, the links between the domestic technological specialisations of national large firms (NLFH), and those of other national firms (ONF), are weak: they are significantly correlated only in Canada and Japan.

Large Firms and Technology: A Case of Non-Globalisation — 71

Finally, the sectoral specialisations of national large firms in foreign countries (NLFA) often reflect those of parent firms (NLFH), with the strong exceptions of France and the USA. In the latter, national large firms are relatively strong in their foreign technological activities in pharmaceuticals, machinery, automobiles, and photography and photocopy, all sectors of relative domestic weakness.

3.6 CONCLUSIONS

The main conclusion of this chapter is that—despite being a critical resource in the global competition and performance of both companies and countries —the production of technology remains far from globalised. Its heavy concentration in the industrialised—as compared to the developing—countries has been recognised for a long time. What we have shown is that, even in the major countries at the world's technological 'core', the production of technology remains highly 'domesticised' in two senses. First, in most of the countries at the world's technological frontier, the foreign technological activities of large firms are still not the major feature. Second, large firms' technological performance is strongly dependent on the performance of the home country, and not independent of it. These conclusions are very similar to the reported results of Porter's recent research on the sources of competitiveness [42]. What happens in home countries still matters greatly in the creation of global technological advantage.

Nonetheless, large firms influence countries in other ways. Large firms are particularly important for the production of technology in R&D-intensive sectors and automobiles. In all our 11 countries, large national firms have a significant influence on sectoral specialisations, whilst other national firms are significant in seven countries, and foreign large firms in three.

Our evidence is in general consistent with the earlier analyses of Dunning [13], Cantwell [7] and our own [35, 39]: country-specific factors create both the general conditions that determine the volume of technological activities, and the specific inducement mechanisms that determine their direction. These lead to accumulated firm-specific advantages that are reflected in international patterns of trade, production and related technological activities. It therefore becomes important to understand the nature of the country-specific factors that make up what Andersen and Lundvall [3] have called 'national systems of innovation', including the system of education, training and basic research that forms the infrastructure for firm-specific technological accumulation.

We also need a better understanding of the reasons why large firms keep most of their technological activities at home. Certain key features related to major innovations may help explain the advantages of geographical concentration: the primacy of multidisciplinary and tacit knowledge inputs, and the commercial uncertainties surrounding outputs. Physical proximity facilitates the integration of multidisciplinary knowledge that is tacit and therefore 'person-embodied' rather than 'information-embodied'. It also facilitates the rapid decision-making needed to cope with uncertainty. For this reason, it may well be more efficient to have technological activities nationally concentrated, with international 'listening posts' and adaptive capabilities maintained through small foreign laboratories, frequent international exchanges often involving what are called 'strategic alliances', and proximity to an internationally outward looking system of higher education [9, 20, 29, 43].

In spite of this, we expect to see greater internationalisation of large firms' technological activities in future, not because it is inherently more efficient, but because it is politically necessary. Uneven technological and competitive developments amongst firms and countries create imbalances, tensions and threats of restrictions on entry into foreign markets. Measures to deal with these threats often involve foreign production and related R&D support, and sometimes independently targeted R&D activities.

In this context, the policies of Swedish large firms are revealing. They perform about 30% of their technological activities outside Sweden (see Table 3.5), and Håkanson and Nobel [22] have found that political factors (particularly those related to establishment within European Community countries) have been important in more than 60% of the decisions taken since 1980 to establish R&D activities abroad.

Similar and even more powerful pressures are now on firms in another technologically high performing country, namely Japan, to expand foreign investment and related R&D activities in Europe and North America [19]. However, we can see from Table 3.5 that they have a long way to go before their technological activities become anywhere near 'globalised'. We can also see that Dutch firms have travelled furthest down this route. It would be intriguing to know whether they intend to continue, or—on simple grounds of managerial efficiency—would prefer in an ideal world to turn back.

3.7 REFERENCES

[1] Z. Acs and D. Audretsch (1989), *Small Firms and Technology*, Directorate of Technology Policy, Ministry of Economic Affairs, The Hague.

[2] F. Amesse, C. Desranleau, H. Etemad, Y. Fortier and L. Seguin-Delude (1991), The Individual Inventor in Canada and the Role of Entrepreneurship, *Research Policy*, **20**, (1), 13–27.

[3] E. Andersen, and B-Å. Lundvall (1988), Small National Systems of Innovation facing Technological Revolutions, in Freeman, Christopher and Lundvall, Bengt-Åke (eds), *Small Nations Facing Technological Revolutions*, Pinter, London.

Large Firms and Technology: A Case of Non-Globalisation ——— 73

[4] G. Bertin and S. Wyatt (1988), *Multinationals and Industrial Property: The Control of the World's Technology*, Wheatsheaf, Brighton.
[5] C. Buckley and M. Casson (1988), *The Future of the Multinational Enterprise*, Macmillan, London.
[6] *Business Week* (1990), The Stateless Corporation, May 14, 52–60.
[7] J. Cantwell (1989), *Technological Innovation and Multinational Corporations*, Basil Blackwell, Oxford.
[8] J. Cantwell and C. Hodson (1990), *The Internationalisation of Technological Activity and British Competitiveness: a Review of Some New Evidence* (mimeo), Economics Department, Reading University.
[9] M. Casson (1990), *Global Corporate R&D Strategy: A Systems View* (mimeo), Economics Department, Reading University.
[10] P. Dasgupta and J. Stiglitz(1990), Industrial Structure and the Nature of Innovative Activity, *Economic Journal*, **90**, 266–293.
[11] G. Dosi (1988a), Sources, Procedures and Microeconomic Effects of Innovation, *Journal of Economic Literature*, **26**, 1120–1171.
[12] G. Dosi (1988b), *Institutions and Markets in a Dynamic World*, The Manchester School, **56**, 119–146.
[13] J. Dunning (1980), Towards an Eclectic Theory of International Production: Some Empirical Tests, *Journal of International Business Studies*, Spring/Summer **1**, 9–31.
[14] J. Dunning (1989), *Multinational Enterprises and the Globalisation of Technological Capacity* (mimeo), Economics Department, Reading University.
[15] J. Fagerberg (1987), A Technology Gap Approach to Why Growth Rates Differ, *Research Policy*, **16**, 87–99.
[16] J. Fagerberg (1988), International Competitiveness, *Economic Journal*, **98**, 355–374.
[17] L. Franko (1989), Global Corporate Competition: Who's Winning, Who's Losing, and the R&D Factor as One Reason Why, *Strategic Management Journal*, **10**, 449–474.
[18] C. Freeman (1982), *The Economics of Industrial Innovation*, Pinter, London.
[19] O. Granstrand, S. Sjölander and S. Alänge (1989), Strategic Technology Management Issues in Japanese Manufacturing Industry, *Technology Analysis and Strategic Management*, **1**, 259–272.
[20] O. Granstrand and S. Sjölander (1990), Managing Technology in Multi-Technology Corporations, *Research Policy*, **19**, 35–60.
[21] Z. Griliches (ed.) (1983), *R and D, Patents and Productivity*, Chicago University Press, Chicago.
[22] L. Håkanson and R. Nobel (1989), *Overseas Research and Development in Swedish Multinationals*, Institute of International Business (RP 89/3), Stockholm School of Economics.
[23] P. Hanson and K. Pavitt (1987), *The Comparative Economics of Research, Development and Innovation in East and West: A Survey*, Fundamentals of Pure and Applied Economics, No.25, Harwood Academic Publishers, Chur.
[24] A. Kleinknecht (1987), Measuring R&D in Small Firms: How Much are we Missing? *Journal of Industrial Economics*, **36**, 253–256.
[25] R. Levin, A. Klevorick, R. Nelson and S. Winter (1987), Appropriating the Returns from Industrial Research and Development, *Brookings Papers on Economic Activity*, **3**, 783–831.
[26] R. Levin, W. Cohen and D. Mowery (1985), R and D, Appropriability, Opportunity, and Market Structure: New Evidence on the Schumpeterian Hypothesis, *American Economic Review*, **75**, 20–24.
[27] F. Malerba and L. Orsenigo (1988), *Technological Regimes, Patterns of Innovation and Firm Variety: a Theoretical and Empirical Investigation of the Italian Case*, (mimeo), Institute of Political Economy, Bocconi University, Milan.

[28] F. Malerba and L. Orsenigo (1990), Personal Communication.
[29] D. Mowery (ed.) (1988), *International Collaborative Ventures in US Manufacturing*, Ballinger, Cambridge, Mass.
[30] R. Nelson and S. Winter (1982), *An Evolutionary Theory of Economic Change*, Belknap, Cambridge, Mass.
[31] A. Pakes and Z. Griliches (1983), Patents and R and D at the Firm Level: A First Look, in Z. Griliches (ed.), *Patents and Productivity*.
[32] A. Pakes and M. Schankerman (1983), The Rate of Obsolescence of Knowledge, Research Gestation Lags and the Private Rate of Return to Research Resources, in Z. Griliches (ed.), *Patents and Productivity*.
[33] P. Patel and K. Pavitt (1987a), Is Western Europe losing the Technological Race? *Research Policy*, **16**, 59–85.
[34] P. Patel and K. Pavitt (1987b), The Elements of British Technological Competitiveness, *National Institute Economic Review*, November, **122**, 72–83.
[35] P. Patel and K. Pavitt (1988), The International Distribution and Determinants of Technological Activities, *Oxford Review of Economic Policy*, Winter, **4**, 35–55.
[36] P. Patel and K. Pavitt (1989), A Comparison of Technological Activities in FR Germany and the UK, *National Westminster Bank Quarterly Review*, May 27–42.
[37] P. Patel and K. Pavitt (1990), Large Firms in Western Europe's Technological Competitiveness, in Mattsson, L-G. and Stymne, B. (eds), *Corporate and Industry Strategies for Europe*, North Holland, Amsterdam.
[38] K. Pavitt (1988a), Uses and Abuses of Patent Statistics, in van Raan, Anthony (ed.), *Handbook of Quantitative Studies of Science and Technology*, Elsevier, Amsterdam.
[39] K. Pavitt (1988b), International Patterns of Technological Accumulation, in Hood, Neil and Vahlne, Jan-Erik (eds), *Strategies in Global Competition*, Croom Helm, London.
[40] K. Pavitt (1990), What We Know about the Usefulness of Science: the Case for Diversity, in Hague, Douglas (ed.), *The Management of Science*, Macmillan, London.
[41] K. Pavitt, M. Robson and J. Townsend (1987), The Size Distribution of Innovating Firms in the UK, 1945–83, *The Journal of Industrial Economics*, **35**, 297–316.
[42] M. Porter (1990), *The Competitive Advantage of Nations*, Macmillan, London.
[43] R. Pearce (1990), Communication at Seminar at Economics Department, Reading University, 9 January.
[44] N. Rosenberg and C. Frischtak (eds) (1985), *International Technology Transfer: Concepts, Measures and Comparisons*, Praeger, New York.
[45] L. Soete (1978), *Inventive Activity, Industrial Organisation and International Trade*, D.Phil thesis, University of Sussex.
[46] L. Soete (1981), A General Test of Technological Gap Trade Theory, *Review of World Economics*, **117**, 638–666.
[47] L. Soete (1987), The Impact of Technological Innovation on International Trade Patterns: The Evidence Reconsidered, *Research Policy*, **16**, 101–130.
[48] L. Soete and S. Wyatt (1983), The Use of Foreign Patenting as an Internationally Comparable Science and Technology Output Indicator, *Scientometrics*, **5**, 31–54.
[49] D. Teece (1986), Profiting from Technological Innovation: Implications for Integration, Elaboration, Licensing and Public Policy, *Research Policy*, **15**, 285–305.
[50] R. Vernon (1966), International Investment and International Trade in the Product Cycle, *Quarterly Journal of Economics*, **80**, 190–207.
[51] R. Vernon (1979), The Product-Cycle Hypothesis in a New International Environment, *Oxford Bulletin of Economics and Statistics*, **41**, 255–267.

Chapter 4

The Internationalisation of Technological Activity and its Implications for Competitiveness

JOHN CANTWELL

An important aspect of the substantial recent extension in the international division of labour within multinational firms has been the specialisation of corporate technological activity across locations, in which each location provides access to specific capabilities complementary to those found elsewhere. This is the major effect of the current reorganisation of international research networks. Using data on the US patenting of the world's largest firms, it is shown that the internationalisation of technological activity is greatest for multinationals based in Britain and the smaller European countries, and in the pharmaceutical and food product sectors. Where multinationals follow a strategy of the international specialisation of technological activity they are likely to rely most heavily on centres of excellence in their industry, but in a few sectors (such as aircraft) locational specialisation and a focus on activity in the main centres are inhibited by barriers to cross-border integration. The forms of interaction between company strategies and the competitiveness of locations vary. At one extreme the movement of activities away from a location simply reflects its competitive decline (the British motor vehicles case); at the other extreme the shift of certain activities brings about a greater specialisation in line with the pattern of technological comparative advantage (the British pharmaceuticals and textiles cases). While countries have tended to narrow their technological specialisation as a result of the new multinational strategies, the major firms have tended to broaden the extent of their technological specialisation. This is illustrated for the leading firms in the pharmaceuticals sector.

Technology Management and International Business: Internationalization of R&D and Technology.
Edited by O. Granstrand, L. Håkanson and S. Sjölander. © 1992 John Wiley & Sons Ltd

4.1 INTRODUCTION

It is widely held that one of the major benefits of multinationality—if not the principal benefit—is the capacity it provides to organise an international division of labour within the firm. Indeed, it seems that this facility has become steadily more important to multinational corporations (MNCs) over the last 20 years or so. MNCs have increasingly tended to rationalise their investments into globally integrated networks [15,6].

Clearly, this has implications for the competitiveness of both firms and countries. Amongst firms there may be pressure to become more multinational and to reorganise any existing network of international operations. The hypothesis that newer or less multinational firms attempt to 'catch up' with the more established MNCs has been investigated elsewhere [5].

At the level of countries those parts of local industry which are heavily influenced by MNC activity may find that the composition of local production moves in new directions. Foreign-owned affiliates may acquire a more specialised role, shifting away from the servicing of local markets and towards a greater involvement in intra-firm trade with other parts of their international networks. MNCs may concentrate research-related production in certain locations and assembly types of production in others, especially within an economically integrated region such as the European Economic Community [2]. The production of indigenous firms which maintain contractual and other links with MNCs may be upgraded or downgraded accordingly. The growing interdependence between production locations and the potential to polarise research-related and other types of production help to explain the attention which is currently paid to 'competitiveness' by governments, policy makers and others.

Given the significance of research to sustaining competitiveness for the firm, and the local presence of research-related production associated with higher value added activities for countries, it is pertinent to examine trends in the international location of research and development more closely. The internationalisation of research is obviously affected by (and affects) the organisational form of the MNC. Until recently it was common to view the MNC as comprising a parent company and a system of affiliates which were largely independent of one another. In this event the degree of geographical dispersion of R&D depends upon the relative strength of competing centripetal and centrifugal forces (for a review see [13]).

The main centralising force is the existence of scale economies in research. The main decentralising force is the need for adaptation and a capacity to respond to local demand-side opportunities as reflected in local markets. Faced with these conditions a typical choice would be to concentrate the bulk of basic or fundamental research in the parent company, while decen-

Internationalisation, Technology and Competition 77

tralising adaptive research more widely. However, matters become rather more complicated once firms adopt strategies of technological specialisation across affiliates, in the same way that they specialise in their productive operations.

There are two reasons why MNCs may take such an internationally integrated approach to technological development. Firstly, technological activity in any industry is locationally differentiated, as part of different national systems of innovation [9]. The distinct characteristics of innovations in each country provide MNCs with an incentive to disperse research facilities to gain access to complementary paths of technological development, which they can then integrate at a corporate level. Secondly, it follows that the geographical dispersion of research to gain access to new lines of innovation may be related to technological diversification. It appears that in recent years there has been a growing interrelatedness between formerly separate technologies, and so to improve technological development even in its own immediate primary field of interest the firm may be obliged to broaden its technological activity through an international strategy.

In this case the emphasis shifts in discussions of the determinants of the internationalisation of technological activity and its consequences. The chief concern is with the supply-side characteristics of locations (their potential for local technological development) rather than their demand-side characteristics (the degree of product adaptation required by local markets). This chapter examines some new evidence on the internationalisation of technological activity in this context. Data on the internationalisation of technological activity by the world's largest firms are related to evidence on the wider geographical composition of innovation in each major manufacturing industry.

Section 4.2 begins by describing the database employed. Section 4.3 uses this to examine the overall extent of the phenomenon of the internationalisation of technological activity, by presenting evidence on its geographical and industrial composition. In the remainder of the chapter, certain issues raised by the emergence of internationally integrated strategies for technological development are explored. Section 4.4 looks at whether MNCs are drawn to locate technological activity in the major centres of excellence for their industry, with specialised expertise in fields complementary to their own. Section 4.5 brings together some of the implications for the technological competitiveness of firms and countries, through an assessment of the relationship between the growth of foreign-owned technological activity within a country and the international strategies of that country's firms. For the purposes of illustration, the focus is on the internationalisation of technological activity by British firms, in the light of the role of the UK as a research centre. Section 4.6 asks whether the internationalisation of technological activity has been associated with a broadening specialisation in the leading MNCs, as a means of exploiting the interrelatedness between different types of skill and technical experience across countries. The pharmaceutical industry is taken as a case study. Some conclusions are set out in Section 4.7.

4.2 THE DATA: HOW PATENT STATISTICS CAN BE USED

The data on technological activity presented here relate to patents granted in the USA between 1963 and 1986. These data are extremely valuable because of the variety of information provided on each patent granted. It is possible to separately identify the firm to which the patent has been granted, the location of the research facility originally responsible for the innovation, and a sectoral classification of the technological activity with which the patent is associated. Through cooperation between researchers at the University of Reading and the Science Policy Research Unit of the University of Sussex it has also proved possible to establish the ultimate ownership of patents where they are granted to affiliates of MNCs.

A consolidation of the technological activity of international corporate groups has been carried out for the world's largest 792 industrial firms. Of these, patenting activity was recorded for 729 firms during the period 1969–86. In these cases in addition to the information just mentioned the nationality of the parent company and the industrial composition of the output of the corporate group were known. The data on this group of firms have been described by Dunning and Pearce [7]. Together, they account for nearly 43% of all patents granted in the USA between 1969 and 1986. Patel and Pavitt [11] discuss further the significance of US patenting by this group of firms relative to all firms.

Patel and Pavitt [11] also elaborate upon the strengths and weaknesses of the use of US patent data in this case. It offers the best (and in many cases the only) source of information on the sectoral and geographical composition of innovative activity across firms. Other recent studies have also supported the suitability of patent data in cross-firm empirical investigations. Griliches, Pakes and Hall [8] argue that despite certain difficulties in using patents to describe time-series trends in innovation, patenting provides a good indicator of cross-firm variation in inventive activity, which is well correlated with R&D expenditure especially for large firms. Acs and Audretsch [1] find that patenting varies across firms with R&D, skilled labour and size in the same manner as innovative activity when the latter is directly measured. Although there are differences in the relationship between patenting and measures of the appropriability of technology (such as concentration in the relevant industry) across firms, Acs and Audretsch

Internationalisation, Technology and Competition ——————— 79

conclude that overall patenting serves as a good proxy for innovative activity. In another survey, Pavitt [12] summarises evidence in favour of using patenting in a third country when making international comparisons of technological activities, and particularly patenting in the USA which enforces common screening procedures and in which large firms generally patent first after their own home country.

There is, however, a drawback to the procedure used for consolidating the patents granted to corporate groups. This arises due to the regular changes in ownership of firms through mergers and acquisitions. To make the task manageable, it was necessary to identify corporate groups at a particular point in time, namely the year 1984. The patent data for the entire period 1969–86 were then consolidated on this basis. Consequently, changes in ownership links during the period were not allowed for.

This means that unfortunately it is difficult to assess the true extent of any trend over time towards the internationalisation of technological activity. Any such trend is likely to be understated in the data on the patenting of the world's largest firms for two reasons. Firstly, where this internationalisation is achieved through acquisition this is not recorded as a change in the geographical composition of the firm's technological development since the affiliate has been considered as part of the corporate group at both the beginning and the end of the period. A study of the foreign-located research of Swedish firms conducted by Håkanson suggested that around 60% of the personnel they employed in foreign R&D facilities worked for affiliates which had been acquired by the parent firm. Secondly, where acquisitions have had motives other than the extension of research facilities (and there have been many of these), it may be expected that the new parent company would tend to wind down affiliate research. Any duplication with the existing research of the MNC may be eliminated, and other functions may be centralised in the technological headquarters. This would appear in the data as a move away from the internationalisation of technological activity.

4.3 THE INTERNATIONALISATION OF TECHNOLOGICAL ACTIVITY AMONGST THE WORLD'S LARGEST INDUSTRIAL FIRMS; ITS GEOGRAPHICAL AND SECTORAL COMPOSITION

The evidence of Table 4.1 must be viewed in this light. Overall, it shows that the world's largest firms witnessed a mild trend towards the internationalisation of technological activity over the 1969–86 period, even without allowing for the effects of acquisitions. The share of US patents granted to

Table 4.1 The share of US patents of the world's largest firms attributable to research in foreign locations (outside the home county of the parent company), organised by the industrial group of parent firms, 1969–86 (%)

	1969–72	1983–86
Food products	16.63	23.96
Chemicals (not elsewhere spec.)	12.18	12.73
Pharmaceuticals	17.69	17.79
Non-metallic mineral products	11.38	15.43
Coal and petroleum products	14.87	13.22
TOTAL	9.80	10.63

Source: The data on the geographical origins and industrial distribution of patents granted in the USA have been compiled at the University of Reading using data on patent counts obtained through the Science Policy Research Unit at the University of Sussex. The data on US patents counts were prepared by the Office of Technology Assessment and Forecast, US Patent and Trademark Office, with the support of the Science Indicators Unit, US National Science Foundation. The opinions expressed in this paper are those of the author, and do not necessarily reflect the views of the Patent and Trademark Office or the National Science Foundation.

these firms attributable to research in foreign locations (outside the home country of the parent firm) rose from 9.8% in 1969–72 to 10.6% in 1983–86.

Identifying industrial groups of firms separately is also revealing. Firms involved in the manufacture of food products, chemicals, pharmaceuticals, non-metallic mineral products (building materials) and coal and petroleum products were especially prone to international research strategies throughout the period. Of these, food and construction material companies experienced a clear trend towards the greater internationalisation of technological activity during the period. By the mid-1980s about a quarter of the innovative activity of the major food firms was located outside their home countries. At the other extreme technological activity is most concentrated in the aircraft and aerospace sector, although it does seem to have become gradually more dispersed since around 1970.

It is also possible to look at the same data classified by the sectoral composition of the technological activity itself, and this is done for comparative purposes in Table 4.2. What emerges most clearly is that the strong internationalisation of research by food product, non-metallic mineral product and coal and petroleum product firms is not especially concentrated in those technological fields in which they are most immediately interested, but in other (presumably related) areas. Equally, although the major textile and instrument firms are not very highly internationalised in their overall technological development, textiles and instruments technologies themselves are rather more subject to the international dispersion of research.

Table 4.2 The share of US patents of the world's largest firms attributable to research in foreign locations (outside the home county of the parent company), classified by technological activity, 1969–86 (%)

	1969–72	1983–86
Food products	12.26	16.73
Chemicals (not elsewhere spec.)	11.85	11.17
Pharmaceuticals	16.27	20.52
Textiles	10.94	11.25
Non-metallic mineral products	7.12	8.64
Coal and petroleum products	8.22	8.18
Professional and scientific instruments	9.44	9.02
TOTAL	9.80	10.63

Source: As for Table 4.1.

Table 4.3 shows that US and Japanese firms are more inclined to concentrate their technological activity in their home country than are firms originating from other industrialised countries. However, whereas US firms have internationalised their research over the 1969–86 period, Japanese companies (in common with the Italian and Canadian groups) have moved towards a more centralised research strategy.

British firms are amongst the most multinational in their organisation of technological activity. Around 40% of the patents they are granted in the USA are attributable to research facilities located outside the UK. This is a much greater share than for their German or French competitors, which acquire less than 15% of their US patents through foreign research. It is higher even than the equivalent shares of Swedish or Canadian firms, though roughly on a par with the position of the largest Swiss firms. The greatest internationalisation of technological activity of all has been undertaken by Dutch and Belgian firms. In 1983–86 nearly three-quarters of the patents they were granted in the USA came from research outside their home countries. Patel and Pavitt [10] have also noted the tendency for the firms of smaller industrialised countries to become substantially internationalised, and in a closer examination of the Swedish case they argue that foreign technological activities represent the cumulative extension of experience and skills first established in the home environment.

It is worth considering the other determinants of the internationalisation of technological activity, apart from the nationality and industrial characteristics of the firm. In this respect as well, the patent data seem to mirror the pattern of innovation suggested by R&D data. Similar results can be obtained to those which have been observed in studies of the internationalisation of R&D. This is seen from a cross-firm regression of the share of patenting attributable to foreign research (SPFR) on an index of techno-

Table 4.3 The share of US patents of the world's largest firms attributable to research in foreign locations (outside the home county of the parent company), organised by the nationality of parent firms, 1969–86 (%)

	1969–72	1983–86
USA	4.28	7.40
Japan	2.85	1.24
Germany	13.57	14.43
UK	43.27	44.91
Italy	20.11	11.72
France	10.23	10.92
Netherlands	63.93	70.04
Belgium	49.62	71.32
Switzerland	44.94	42.59
Sweden	20.94	31.30
Canada	42.06	35.52
Others[a]	32.76	23.10
TOTAL	9.80	10.63

Source: As for Table 4.1.
[a] Excluding companies registered in Panama.

logical competitiveness of the firm (TC), its foreign production ratio (FPR), and the overall size of its technological activity (STA). The index of technological competitiveness is defined as the firm's share of patenting in its industry in 1969–72 relative to its share of industry sales in 1972 amongst the world's largest firms. An adjustment was made to allow for the higher propensity to patent in the USA of US firms. The foreign production ratio is represented by the sales from international production (foreign affiliate sales less intra-firm exports from the parent company) relative to the global sales of the corporate group. The overall size of technological activity is measured by the total number of US patents granted to the firm in the period in question, divided by a constant factor to make its coefficient easier to handle. The results of the regression run for patenting over the 1969–86 period were as follows, with standard errors in brackets:

$$SPFR = -0.014 + 0.001 \, TC + 0.760^* \, FPR - 0.159^{**} \, STA$$
$$(0.03) (0.01) (0.06) (0.06)$$

No. of observations = 140 $R^2 = 0.533$
F = 51.726 $\bar{R}^2 = 0.523$

* Denotes coefficient significantly different from zero at the 1% level.
** Denotes coefficient significantly different from zero at the 2% level.

Predictably enough, the degree of multinationality of the firm is very strongly related to the extent of internationalisation of its technological activity. A positive association might also have been expected with the technological competitiveness of the firm, but it was not significant. The negative sign of the coefficient on the overall size of technological activity is in line with the study of Pearce [13] using R&D data. He reported that the size of the firm as measured by its sales was negatively related to the extent of decentralisation of R&D, though at a declining rate as size increased. However, although this is consistent with what others have found using evidence on the location of R&D expenditure, it remains somewhat surprising. It seems reasonable to hypothesise the reverse, whether due to the existence of scale economies in R&D or the greater capacity to engage in an international research strategy as the size of technological activity rises. It may be that this is the consequence of the composition of firms in the sample. US and Japanese firms have the highest levels of patenting (between them they account for nearly three-quarters of all patents granted in the USA to the largest firms), but their technological activities are the most highly centralised.

4.4 THE INTERNATIONAL LOCATION OF INNOVATIVE ACTIVITY AND THE TECHNOLOGICAL SPECIALISATION OF HOST COUNTRIES

Largely because of the size of technological activity of US and Japanese firms, and the significance of their home-based operations, the bulk of the research of the world's largest firms is concentrated in the USA or Japan. In 1978–86 55.5% of US patents granted to the largest firms were attributable to research facilities in the USA. Of those attributable to innovations in non-US locations, 44.8% were from Japan, 25.2% from Germany, and 9.1% from the UK. European locations collectively accounted for 52.1% of innovations outside the USA.

This also illustrates the high degree of concentration of the technological activity of the largest firms in the industrial centres. Taking all patents granted in the USA in 1978–84 (not just to this group of firms), of those due to research outside the USA 32.5% were attributable to Japan, 23.0% to Germany, 9.2% to the UK, and 55.3% to the European countries as a group. This left 12.2% due to research outside Europe or Japan, as opposed to 3.1% of patents granted to the biggest firms.

Perhaps more interesting is the comparative advantage of different locations as hosts to innovative activity, in terms of which industrial types of activity they are most likely to attract. It may be hypothesised that if MNCs regard their international technological activity as a source of innovation for the company as a whole, host countries will be most attractive to MNCs in their own fields of national specialisation. To examine this in more detail an index of comparative advantage, or what is termed revealed technological advantage (RTA), has been constructed for the major industrialised countries as locations for innovative activity. This is defined as the share of patents attributable to research in the country in question in a given industry, relative to the share of all patents which can be traced to that country. The index thus varies around unity, with those industries in which the country is especially attractive as a location for research assuming values of the index greater than one. Owing to the differences involved in patenting US as opposed to non-US inventions in the USA, the index is compiled relative to all patents granted for the USA, but relative to all patents attributable to foreign locations for all other countries.

The results of this exercise are reported for selected sectors in the UK and Japanese cases in 1978–86 in Table 4.4. The data are classified by the nature of the underlying technological activity with which each patent is associated, rather than by the industrial characteristics of the firms responsible. The same procedure was repeated for patenting attributable to research in foreign locations for the world's largest firms. In this case the position of each country depends entirely upon the distribution of the innovative activity of foreign-owned firms carried out locally. The index here therefore serves as an indicator of the attractiveness of a particular country for

Table 4.4 The revealed technological advantage of the UK and Japan in overall innovation, and in the foreign research of the world's largest firms (outside the home country of the company), classified by technological activity, 1978–86

	Japan foreign	Japan overall	UK foreign	UK overall
Food products	0.52	0.69	0.78	1.43
Chemicals (not elsewhere spec.)	0.99	0.87	1.13	1.03
Pharmaceuticals	0.82	0.74	2.13	1.72
Metals	0.52	0.74	0.86	1.06
Electrical equipment (not elsewhere spec.)	1.33	1.21	0.73	0.95
Office equipment	0.98	1.80	0.92	0.64
Motor vehicles	3.11	1.54	0.83	0.68
Aircraft	0.00	0.18	0.00	2.10
Textiles	0.00	0.61	1.99	0.94
Professional instruments	1.52	1.44	1.08	0.83

Source: As for Table 4.1.

Internationalisation, Technology and Competition ——————— 85

a given technological activity as a location for the research of foreign multinationals. The international location of innovative activity by the world's largest firms needs to be examined in the context of the wider role of locations as centres for technological development. It was argued above that the major MNCs have an incentive to locate research facilities in all the most important international centres of innovation in their industries.

The choice of the UK and Japan in Table 4.4 is because they represent the two extreme cases. In Japan, foreign MNCs have been especially clearly drawn to locate research locally in the fields of Japanese strength. In the UK this has not always been the case.

The UK as a location seems to have experienced a high degree of mobility in the pattern of its technological advantage between 1963–70 and 1978–84 [3]. This reflects changes in the UK research activities of the world's largest firms. The sectoral composition of both all innovation in the UK and foreign-owned research activity in the UK shifted between the 1960s and 1980s. At the same time, partly due to the effect of switches of foreign research on all innovation in the UK, the relationship between the pattern of foreign-owned and indigenous technological activity changed.

As a rule, for most other major industrialised countries the distribution of foreign-owned research activity was positively related to the overall pattern of innovation, though it was not always significantly so. The UK was an important exception in the earlier 1969-77 period. For the UK at this time there was a significant negative association between the distribution of foreign-owned technological activity and the composition of overall innovation of all UK-located firms and individuals. In other words, in the late 1960s and early 1970s foreign firms were especially prone to carry out research in the UK in areas of technological activity in which the country was comparatively disadvantaged, or at least much less advantaged. The most striking sector is aircraft, in which the revealed technological advantage of the UK was 1.93 overall as against 0.27 in foreign-owned research activity; foreign firms were also less likely to undertake technological projects in the UK than were indigenous concerns in food products, construction and mining equipment, general industrial equipment, rubber and plastics products, non-metallic mineral products, and coal and petroleum products. Their research efforts were relatively stronger, though, especially in textiles, but also in pharmaceuticals, metals, office equipment and motor vehicles.

However, the negative relationship between foreign-owned and overall technological activity in the UK from 1969–77 disappeared once two other considerations were taken into account. Firstly, in sectors in which international production is relatively low only small numbers of patents are granted to foreign-owned research facilities. The problems of calculating the

RTA index where only small numbers of patents are involved are familiar [3]. Essentially, the cross-sectional RTA distribution becomes skewed and its variance rises, while stochastic factors play a greater role in influencing the composition of the distribution. In this instance, particular problems arise in the case of aircraft and nuclear reactors, in which little research (or production) is carried on outside the home country.

Secondly, the extent of the international integration of economic activity discussed above in the opening section varies across industries. Where national markets remain substantially differentiated, or where government regulations and other non-tariff barriers are significant, progress towards integration may be prevented. In such cases, the international rationalisation of technological activity concentrated in the major centres of excellence may not be feasible. Instead, research facilities may be dispersed in large part in accordance with the requirements of local markets (food products) or national procurement policies (pharmaceuticals), or concentrated especially at home where political and business contacts enable contracts to be won (aircraft).

The first factor was allowed for by dropping the two sectors with small numbers of patents from the regression of the sectoral distribution of foreign-owned RTA on the equivalent overall RTA distribution. The second issue was dealt with by the introduction of an additional independent variable which measured the degree of integration of activity achieved by multinationals across industries. This was proxied by the ratio of exports to sales of the foreign affiliates of US multinationals in the relevant sector in 1982, reflecting their international as against their local role. The effect of the first change was to eliminate the significantly negative relationship in the UK in 1969–77. The second change worked in the same direction, but the integration variable did not itself exercise a significant effect on the distribution of foreign-owned research.

In any case, the comparison between foreign-owned and overall technological activity in Britain had turned around by 1978–86. In this later period, as for other countries, there was a positive relationship between the sectoral distribution of foreign-owned firm and overall research in the UK, though it was still not a statistically significant one. It seems that this transformation can be largely explained by changes in two fields of technological activity, namely pharmaceuticals and motor vehicles. In pharmaceuticals foreign firms became still more attracted to locate research in the UK, while local firms also steadily improved their innovative performance; while in motor vehicles foreign firms reduced their UK research in line with indigenous producers. The respective virtuous and vicious circles which characterised the organisation of these two sectors in the UK have been described elsewhere [2].

It remained the case, though, that foreign firms were virtually absent from work on aircraft technology in the UK; they did not carry out any research in the UK in this field which led to US patenting in the years 1978–86. This is despite the strength of UK-owned firms in this area, but as noted in Tables 4.1 and 4.2 it is a field in which research is heavily oriented towards the domestic locations of the companies responsible. In this sector there does not seem to be any particular incentive to locate research in other international centres of innovation; this may be because of its linkage with defence-related R&D in which the pull of home country government contracts is important.

By comparison, the association between the pattern of foreign- and domestically owned research was closest or most significant in Japan, and became stronger still between 1969–77 and 1978–86. Foreign firms are especially keen to invest in research in the major fields of local technological strength in Japan, namely electrical equipment, motor vehicles and professional and scientific instruments. The switching of foreign-located research towards Japan has been most dramatic in the motor vehicles area. In this sector the revealed technological advantage of Japan as a location for foreign-owned research rose from 0.48 in 1969–77 to 3.11 in 1978–86 (see Table 4.4).

In Germany and Switzerland the relationship between foreign and local technological activity is reasonable, but the puzzle is why foreign firms have not been relatively drawn to undertake research in the chemicals and pharmaceuticals fields. Aside from the significance of government regulations in pharmaceuticals which have drawn foreign firms to other countries such as France, the very strength of German firms may have inhibited entry for some time. However, recent trends suggest that this may be changing. Technological activity has been attracted in the bleaching and dyeing fields, especially in the 1980s, which are the areas of the greatest traditional strength of German and Swiss firms. Foreign-owned research has also been pulled towards Germany in the field of general industrial equipment, and to Switzerland in chemical equipment, and more recently in food products, which are all areas of indigenous capability. In Italy foreign firms have shown an increasing propensity to locate technological activity in specialised industrial equipment, in which the country is comparatively advantaged. In France the overlap between foreign and indigenous research is most obvious in the telecommunications field, although foreign firms are also drawn to local technological activity in office and computing equipment in which the overall French record is unimpressive.

On the whole, though, it does seem that the technological activity of the world's largest firms is quite frequently drawn to the main centres of innovation for their industries. In line with the general tendency towards international integration this seems to have become stronger, though in some sectors more than others. Of course, this is an iterative rather than a one-way process, in that the establishment of research by this crucial group of companies itself contributes to the overall technological activity of a location, both directly and indirectly, through the spur that they provide to the local research of other firms, whether competitors or contractual partners.

One question which then arises is whether the companies which have adopted international research strategies to take advantage of the innovative potential of different locations have achieved better overall performance as a result. A first approximation to whether this is so can be achieved through a cross-firm regression of the proportional growth in sales between 1972 and 1982 (PGS) as a measure of performance on the share of patenting attributable to foreign research at the start of the period 1969–72 (SPFR) and the index of technological competitiveness at the beginning of the period as previously described (TC). The results of this regression were as follows, with standard errors in brackets as before:

$$PGS = -0.995^* + 0.143^{**}SPFR + 0.097^*TC$$
$$(0.04) \quad (0.09) \quad (0.02)$$

No. of observations = 320 $R^2 = 0.067$
F = 11.455 $\bar{R}^2 = 0.062$

*Denotes coefficient significantly different from zero at the 1% level.
**Denotes coefficient significantly different from zero at the 15% level.

The share of patenting attributable to foreign research is positively associated with the growth of the firm in the ensuing period, as hypothesised. So firms with a greater international dispersion of research have tended to have a better record of performance. However, the coefficient on SPFR is not significantly different from zero at the statistical levels which are customarily employed. Although the measure of technological competitiveness is positively and significantly related to growth as expected, the overall explanatory power of the equation is quite weak.

One reason for the weakness of this association may be that SPFR is a rather crude measure of the capacity of the firm to capture the benefits of an international research strategy. It takes no account of the actual geographical composition of the technological activity of the firm (the extent to which it is in centres of innovation), nor does it allow for differences in the appropriability of innovations across firms. These matters will be investigated more closely in future research.

4.5 THE INTERACTION BETWEEN THE INTERNATIONALISATION OF TECHNOLOGICAL ACTIVITY AND NATIONAL COMPETITIVENESS: THE CONSEQUENCES FOR BRITISH FIRMS AND INDUSTRIES

From the perspective of the UK, the internationalisation of technological activity can be viewed in two ways. Firstly, there is the role of research carried out abroad by British MNCs. This may improve their competitiveness but not necessarily the performance of their UK operations. Secondly, the UK is host to the research facilities of foreign-owned firms. It has been suggested above that where foreign MNCs locate more extensive fundamental research in Britain this is generally associated with a beneficial impact upon local industry [2]. The share of research-related production tends to rise and export performance to improve, setting in motion a positive interaction with the local research and production of indigenous firms. An alternative argument has recently been advanced by Stoneman [14]. He contends that the research of foreign-owned firms in Britain represents a potential 'internal brain drain' in which the UK's scientific and technological resources are used by foreign MNCs to support production elsewhere in the world, with no spin-off benefits for local British industry.

A full examination of these issues awaits further research, but it is possible to extend the discussion here to survey in rather more detail how the internationalisation of technological activity has impinged upon British firms and industries. Considering British firms first, Table 4.3 has already indicated their relatively heavy dependence upon foreign research facilties. A more detailed treatment of certain of the underlying patterns behind these data is provided in Table 4.5. The British producers of food products, machinery and mechanical equipment and coal and petroleum products are particularly obliged to rely on research outside the UK. However, it seems that a clear trend towards the internationalisation of technological activity has been underway in the motor vehicles, textiles and non-metallic mineral products sectors. It should be noted that for UK textile firms this is true in the textile field, even though other UK firms outside the sector have reduced their share of foreign technological activity in textiles from a very high level. Nonetheless, the rising share of foreign research for UK textile firms is also partly associated with technological diversification, and is a function of a strong internationalisation of technological activity in the field of professional and scientific instruments (the other instruments and controls category). Textiles is indeed the most interesting case, since in motor vehicles and to a lesser extent in non-metallic mineral products there has been a relative

Table 4.5 The share of US patents of the largest UK firms attributable to research outside the UK, organised by the industrial group of parent firms, 1969–86 (%)

	1969–72	1983–86
Motor vehicles	10.99	33.33
Textiles	18.45	34.48
Non-metallic mineral products	12.12	42.11
TOTAL	43.27	44.91

Source: As for Table 4.1.

decline of innovation in the UK. By contrast, for the world's largest firms there has actually been an increase in the comparative importance of Britain as a location for technological activity in the textiles field. In other words, it would seem that whereas in motor vehicles British firms have simply followed the trend and moved their research facilities elsewhere (or closed down local research anyway), in textiles the internationalisation of their research has been part of a much more positive process. Either research abroad has been supportive of an extension of their UK research, or non-UK firms have increased their British research operations. To the extent that this has happened it may be partly in response to the expansion of the technological activity of British companies abroad.

Table 4.6 offers a comparison between the role of foreign-owned research in the UK and in two other large EC countries, Germany and France. Just as British firms are more heavily internationalised in their technological activity than German or French firms (Table 4.3), so too are local British industries relative to German or French. Of US patents attributable to research in each of these three countries in 1983–86, 15.2% were granted to foreign-owned firms in the UK, 8.8% in Germany and 8.9% in France. However, the increasing role of foreign-owned research was evident in all three countries, as can be seen from the equivalent shares of patenting in the earlier 1969–72 period, which were 10.9% in the UK, 5.6% in Germany and 6.3% in France. The sectoral composition of this internationalisation by the type of technological activity involved is also similar across this group of countries, but especially between Britain and France. The share of foreign firms in local innovation is especially high in both Britain and France in chemicals (particularly organic chemicals), pharmaceuticals and office equipment; and it is relatively low in aircraft, other transport equipment and other manufacturing.

Differences arose in the case of textiles, in which foreign-owned technological activity is comparatively low in France but not in the UK, and in

Table 4.6 The share of foreign research in US patents deriving from the three largest European host countries, 1969–86 (%)

	UK	Germany	France
Food products	14.81	10.81	9.52
Chemicals (not elsewhere spec.)	20.33	6.18	13.99
Organic chemicals	25.04	5.61	18.00
Pharmaceuticals	31.48	9.36	17.76
Office equipment	24.45	14.07	21.46
Motor vehicles	14.80	5.63	18.79
Aircraft	0.00	3.28	2.41
Other transport equipment	6.00	7.59	2.14
Textiles	13.85	1.71	2.78
Rubber and plastics products	7.89	6.60	7.78
Other manufacturing	4.85	2.32	2.52
TOTAL	15.15	8.77	8.86

Source: As for Table 4.1.

rubber and plastics products which is the other way round. Foreign firms have also been relatively weakly represented in patenting originating from research in Germany in the textiles field, so here it seems that Britain is the exception rather than the rule. As in the UK or France, foreign-owned research facilties in Germany account for a high share of US patenting in office equipment, and low shares in aircraft and other manufacturing. By contrast, though, foreign firms are especially prominent in technological activity in the coal and petroleum products field in Germany.

It may also be recalled from Table 4.5 that UK firms had increased the internationalisation of their technological activity in the textiles and non-metallic mineral products industries, despite an improvement in the relative share of large firm technological activity in textiles located in the UK over the same period. In textiles the foreign firm share of British research has not altered very much. In terms of the activity of the largest firms the foreign share in the textiles field has fallen, suggesting that more British textile company-owned research outside the UK (in a variety of technological areas) has been associated with an extension in the domestic research facilities of British textiles firms in the textiles field. In non-metallic mineral products the foreign share has risen slightly at the industry level, and has risen significantly in the building materials technological field. This may indicate some withdrawal from the UK by indigenous companies in non-metallic mineral products (like in motor vehicles), but it may also show some response on the part of foreign firms to the internationalisation of technological activity by the largest UK firms.

4.6 CHANGES IN THE TECHNOLOGICAL SPECIALISATION OF MNCs

The relationship between the internationalisation of technological activity and the competitiveness of countries may vary, as the contrasting experience of research in motor vehicles and textiles in the UK indicates. Centres which start from a position of comparative technological advantage may make cumulative gains at the expense of others. In the UK such gains have also been observed in the pharmaceuticals sector, in which Britain, the USA, Germany and Switzerland constitute the major centres of excellence [2].

The same developments can be viewed from the perspective of firms rather than countries. Returning to an earlier theme, MNCs invest in technological activity in the main centres for their industry to take advantage of locationally differentiated expertise. By specialising in accordance with local strengths in each centre companies aim to broaden their profile of interrelated activities which can then be integrated at a corporate level. If MNCs have been restructuring their international technological activity in this fashion, it can be hypothesised that this process will tend to have been associated with a broadening out of their technological specialisation. This would be the reverse of what would be expected at a country level, as national patterns of technological specialisation are reinforced and thus tend to become narrower. There has indeed been a narrowing in the degree of technological specialisation of most countries between the 1960s and 1980s [3], though this locational agglomeration in comparative technological advantage is due to other factors as well as the internationalistion of research [4].

Just as there is evidence that countries are becoming more specialised, the data on the world's largest firms suggest that they are broadening out their portfolios of technological activity, though into related not unrelated areas. Again, the internationalisation of technological activity is not the only explanation of this trend, but it has played a part. The tendency for the technological specialisation of companies to widen can be illustrated in the case of the pharmaceuticals industry, for firms from the leading centres (Table 4.7).

The measure of the degree of technological specialisation used is the variance of the cross-sectoral distribution of shares of US patenting over 33 areas of technological activity. The variance is very high where technological activity is concentrated in a single sector (say, 80% of patents are in the pharmaceutical field), but it is low where activity is spread over a broader range of sectors.

Table 4.7 shows that US, German and Swiss pharmaceutical firms have become more similar in the extent of their technological specialisation. That is, they have all moved towards a broad (and away from a narrow) specialisation both at home and abroad. They have especially broadened out their

Table 4.7 The degree of technological specialisation of pharmaceutical firms by the nationality of parent company, measured by the variance of the cross-sectoral distribution of their US patents, 1969–86

	Domestic research	
	1969–72	1983–86
USA	88.49	56.85
Germany	136.35	76.18
UK	272.33	318.31
Switzerland	256.28	92.99

	Foreign research	
	1969–72	1983–86
USA	138.81	81.67
Germany	68.75	76.93
UK	150.51	159.89
Switzerland	250.09	130.56

Source: As for Table 4.1.

domestic technological activity, to a point where they are now more focused in their foreign research than they are at home. This might have been expected, if the internationalisation of technological activity comes to be viewed as providing a more specialised complement to efforts in the home centre. However, for US and Swiss firms the pattern of foreign technological activity was itself also widened, and the firms of all these three countries are not much more specialised abroad than at home. In other words, their international activities in other centres also cover investments in a reasonable spectrum of research fields.

The exception is provided by British firms. Not only do they remain highly specialised in their technological activity, but they have become more so. Indeed, they are still more heavily specialised in the UK than they are abroad. The explanation has to do with patterns of comparative advantage in technological activity as between national groups of firms, and the nature of shifts in the composition of technological opportunities across fields. UK firms have specialised in fields in which technological opportunities have risen most, causing them to consolidate their existing pattern of specialisation. As a result of changing opportunities, all firms have shifted away from research related to industrial chemicals and towards pharmaceuticals and biotechnology. British companies already had a comparative specialisation in these medicinal and biologically related technological fields, which has now been enhanced. This is especially true of their research in the UK itself. So in this case although foreign operations are also specialised in pharmaceuticals and biologically based technologies reflecting the relative strength of UK MNCs in the sector, foreign research provides a broader balance to support the more concentrated efforts of the British parent companies.

Overall, there is a trend towards a broadening degree of technological specialisation among the leading MNCs, though they are moving into related and hence complementary technological fields. In the pharmaceuticals industry, US and German firms have moved towards research in pharmaceuticals technology itself, while Swiss firms have also increased the share of their activity in chemical processes. However, trends in specialisation are subject to shifts in the pattern of technological opportunities, and depend on the comparative advantage of firms. Where companies are comparatively advantaged in fields of greater opportunity their initial specialisation is likely to be consolidated.

4.7 CONCLUSION

A wide range of aspects of a useful source of new evidence on the internationalisation of technological activity has been reviewed here. It can be concluded that this is an important phenomenon which deserves closer attention than it has typically been paid in the existing literature. One impression that clearly emerges from the data is that previous writers may well have understated the role of the internationalistion of research due to the reliance of most work in this field on evidence from US companies alone.

In general, it seems that US and Japanese MNCs have made comparatively little use of international research strategies by the standards of European firms. However, it also appears that US firms have begun to appreciate the benefits of a wider dispersion of technological activity, and they have increasingly made use of foreign research facilities. Where the world's largest industrial firms have dispersed their research they have as a rule been especially attracted to the main centres of innovation for their primary sector of activity, although there are exceptions, as in the aircraft industry.

British firms and British industry have both been heavily involved in the internationalisation of technological activity, as have the firms and industries of the smaller European countries. In pharmaceuticals and perhaps in textiles in the UK case this has been associated with an improvement in competitive performance, while in motor vehicles there has been a competitive deterioration. It can be seen that the internationalisation of technological activity affects the competitiveness of centres in different ways, in which in some industries certain locations tend to gain at the expense of others. However, overall it is possible that an increased technological specialisation

Internationalisation, Technology and Competition ─────────── 95

of countries will develop which may be mutually beneficial, each enhancing their specific areas of expertise. Both foreign owned research in a country and the international activities of domestically based firms have tended to have the effect of increasing national specialisation.

Within the international networks of MNCs, however, technological activity is becoming broader. This is a means of exploiting the advantages of specialisation in affiliates in accordance with the location-specific characteristics of technology, together with the advantages of conducting a range of increasingly interrelated technological activities. The internationalisation of technological activity has become the basis for constructing a cross-country network of facilities geared to technology creation.

4.8 REFERENCES

[1] Z.J. Acs and D.B. Audretsch (1989), Patents as a measure of innovative activity, Wissenschaftszentrum Berlin für Sozialforschung Discussion Paper, FS IV 89–5.
[2] J.A. Cantwell (1987), The reorganisation of European industries after integration: selected evidence on the role of multinational enterprise activities, *Journal of Common Market Studies*, **26**, No. 2, December.
[3] J.A. Cantwell (1989), *Technological Innovation and Multinational Corporations*, Basil Blackwell, Oxford.
[4] J.A. Cantwell (1991), The international agglomeration of R&D, in Casson, M.C. (ed.), *Global Research Strategy and International Competitiveness*, Basil Blackwell, Oxford.
[5] J.A. Cantwell and F. Sanna Randaccio (1990), The growth of multinationals and the catching up effect, *Economic Notes*, **19**, No. 1.
[6] J.H. Dunning (1988), *Explaining International Production*, Unwin Hyman, London.
[7] J.H. Dunning and R.D. Pearce (1985), *The World's Largest Industrial Enterprises, 1962–1983*, Gower, Farnborough.
[8] Z. Griliches, A. Pakes and B.H. Hall (1987), The value of patents as indicators of inventive activity, in Dasgupta, P. and Stoneman, P. (eds), *Economic Policy and Technological Performance*, Cambridge University Press, Cambridge.
[9] B.A. Lundvall (1988), Innovation as an interactive process: from user-producer interaction to the national system of innovation in Dosi, G., Freeman, C., Nelson, R., Silverberg, G. and Soete, L.L.G. (eds), *Technical Change and Economic Theory*, Pinter, London.
[10] P. Patel and K. Pavitt (1988), Large firms in Western Europe's technological competitiveness, paper presented at the Prince Bertil Symposium on Corporate and Industrial Strategies for Europe, Stockholm, November.
[11] P. Patel and K. Pavitt (1989), Do large firms control the world's technology?, *University of Sussex Science Policy Research Unit Discussion Paper*, January.
[12] K. Pavitt (1987), Uses and abuses of patent statistics, *University of Sussex Science Policy Research Unit DRC Occasional Paper*, No. 41, February.
[13] R.D. Pearce ((1989), *The Internationalisation of Research and Development by Multinational Enterprises*, Macmillan, London.
[14] P. Stoneman (1989), Overseas financing for industrial R & D in the UK, paper presented at the Annual Meeting of the British Association for the Advancement of Science, Sheffield, September.
[15] R. Vernon (1979), The product cycle hypothesis in the new international environment, *Oxford Bulletin of Economics and Statistics*, **41**, No. 4, November.

Chapter 5

Locational Determinants of Foreign R&D in Swedish Multinationals

LARS HÅKANSON

Drawing on detailed data on some 150 foreign R&D establishments belonging to the 20 largest manufacturing enterprises in Sweden, this chapter develops and tests a number of hypotheses as to the locational determinants of different types of such units. Although, as expected, locational determinants differ between different kinds of foreign R&D activities, in general, 'demand-related' factors—market potential and need for market adaptation in the US—are clearly more important than 'supply-related' factors associated with high host country R&D intensity. Average age of subsidiaries and ease of intra-group communication, associated with 'psychic distance' from Sweden, appear to be significant determinants for most types of foreign R&D. The location of foreign R&D facilities seems also to be influenced by 'political' factors; Swedish MNCs perform a larger share of their foreign R&D in the six original EEC countries than can be explained by the size of their respective markets and their relative proximity to Sweden. However, foreign R&D locations also mirror essentially 'random' factors: 'historical accident' in the form of R&D unrelated acquisitions, the activities of unusually entrepreneurial subsidiary managers, etc.

5.1 INTRODUCTION[1]

Since the mid 1980s, internationalization of research and development (R&D) conducted by multinational companies (MNCs) has become an important item on many business executives' agenda. Concurrently, geographical and

---
[1] Valuable comments on an earlier draft were kindly provided by Robert Nobel, Ivo Zander and Udo Zander and are gratefully acknowledged.

Technology Management and International Business: Internationalization of R&D and Technology.
Edited by O. Granstrand, L. Håkanson and S. Sjölander. © 1992 John Wiley & Sons Ltd

organizational decentralization of R&D in multinational companies has increasingly begun to attract academic attention [1, 6, 23, 24, 33].

A number of recent studies convincingly demonstrate the absolute and relative significance of foreign R&D, especially in European multinationals. In the sample of 20 leading Swedish MNCs analyzed in this chapter [12, 13, 14], foreign R&D in 1988 accounted for about 23% of the total. In a study of 23 German multinationals, foreign R&D personnel numbered 16 000, 17% of total R&D employment in these companies [33]. In an analysis of US patent data, Patel and Pavitt [23] show that foreign subsidiaries account for substantial shares of the patent applications filed by multinational groups headquartered in smaller European countries such as Sweden, the Netherlands and Switzerland.

R&D undertaken in foreign-owned subsidiaries is sometimes seen as a form of disguised 'brain-drain', i.e. when engineers and scientists employed are in scarce supply and work on problems of little relevance to the local economy. However, most governments tend to be favorably disposed to hosting MNC R&D. Subsidiaries performing R&D, it is argued, are more likely to be engaged in technologically advanced production, in turn assumed to be associated with job creation, exports and a generally favorable impact on domestic industry.

However, activities labeled 'R&D' include a wide range of different functions. The tasks of overseas R&D units range from relatively routine adaptations of products and processes, over new product development to long range research [11, 25, 26, 27]. The present chapter proceeds on the assumption that the locational characteristics of each type of activity differ, reflecting a wide range of different factors and considerations. These include the nature and intensity of the linkages maintained to other corporate functions, customers and other research units, as well as environmental conditions, such as labor market conditions for technicians and scientists, government policies and the nature of local demand. Whereas a companion paper [14] analyzes the influence of various *company characteristics* on the relative volumes of different kinds of R&D performed abroad, the present chapter focuses on the influence of *locational characteristics* on the performance of different types of foreign R&D activities.

In the literature, the observed internationalization of R&D is often related to an assumed need to 'tap into' foreign sources of technical competence and development. Over time, 'supply-related' factors—the availability of highly skilled scientists and engineers, the presence of a dynamic 'scientific infrastructure', etc.—are seen to have become more important than 'demand-related' ones, such as the need for customer contact and market proximity in order to adapt products and processes [1]. In other words, the impetus for foreign R&D is increasingly the need to 'know how' rather than 'know what' to develop [6].

Foreign R&D in Swedish Multinationals 99

Hypotheses or statements of this type are only partially supported by empirical analyses of individual company cases [15, 16, 33], in which existing locational patterns of foreign R&D establishments must often be interpreted as the result of 'historical coincidence', e.g. acquisitions undertaken for reasons often totally unrelated to R&D, the efforts of entrepreneurial subsidiary managers, etc. Similarly, organizational structures and other company-specific variables seem to have a strong influence on companies' willingness to carry out R&D at foreign locations [14].

The strive for profit maximization, as enforced by 'the discipline of the stock market' is sometimes assumed to lead to some sort of optimal location pattern of foreign R&D establishments. However, even if such a long term optimum should exist (and be attainable within the organizational and cognitive limits of real world companies), there is no reason to believe that it would correspond very closely to the locational pattern existing at any one time in any one individual corporation.

In view of the above considerations, observed locational distributions of foreign R&D activities cannot be expected to very closely resemble those that would be prescribed by some theoretical 'optimum'. Thus, the primary aim of the hypotheses advanced and of the statistical models tested is not to 'explain' the locational patterns observed. The object is (1) to test the hypothesis that the locational patterns of different types of foreign R&D units are influenced by different locational factors, and (2) to identify a number of such factors.

The next section gives a brief description of the database used in the empirical analysis. Section 5.3 outlines a typology of different kinds of overseas R&D units. Following, in Section 5.4, the formulation of a number of hypotheses as to the influence of various country characteristics on the location of different types of overseas R&D units in Swedish multinationals, Section 5.5 presents the statistical tests and their results.

5.2 THE DATABASE

The bulk of the data used in this study was collected by means of a questionnaire administered to the 20 largest Swedish manufacturing enterprises in the chemical and engineering industries[2]. In 1987, these 20 firms spent a total of 11.9 billion Swedish crowns (approx. US $ 1.9 billion) and

---

[2] AGA, Ericsson, Perstorp, SKF, Alfa-Laval, ESAB, Pharmacia, Swedish Match, Astra, Fläkt, PLM, Tetra Pak, Atlas Copco, Kantal Höganäs, Saab-Scania, Trelleborg, Electrolux, Nobel Industrier, Sandvik, and Volvo. Formally, Tetra Pak is not a Swedish company but its R&D is mainly performed in Sweden. Asea Brown Boveri (ABB) was not included, largely because it was judged unlikely that company officials would be able to find time to complete a lengthy questionnaire during the restructuring following the merger between Asea and BBC.

employed some 24000 persons in R&D in Sweden, thereby accounting for between two-thirds and three-quarters of all R&D in Swedish industry. The questionnaire identifies some 170 foreign R&D laboratories, belonging to these companies[3]. With a total employment of around 8 100 people, these foreign units account for close to one-quarter of the total R&D effort of the 20 companies[4].

The administration of the questionnaire was tailored to fit the organizational structures and reporting systems of each company. In 15 companies, it was completed by the corporate R&D department, either on the basis of centrally kept records (nine cases), or following extensive consultation with managers at divisions and/or major subsidiaries (six cases). In a further five companies, questionnaires were distributed and data collected by the researchers, based on lists of appropriate officials provided by headquarters staff.

The questionnaire consists of two parts. The first part ('A-form') was designed to obtain general and background information. The second part of the questionnaire ('B-form') was completed for each foreign R&D department belonging to the unit in question. A summary of the answers from the questionnaires was sent to the manager responsible for R&D at corporate headquarters in order to check for plausibility, and to help clear out uncertainties. It is likely that the quality and reliability of the data vary with the different ways in which they were obtained. However, data collection was designed to ensure—as far as possible—that questionnaires be answered by officials in the company directly involved with the questions asked[5].

5.3 A TYPOLOGY OF FOREIGN R&D UNITS

Several typologies of foreign R&D activity have been suggested in the literature [4, 5, 11, 26, 27]. A recent synthesis is provided by Pearce [25], classifying laboratories on the basis of the tasks they perform and the types of organizational linkages they maintain. Here, we shall draw on an empirically generated classification—by and large compatible with Pearce's typology—emphasizing the *motives* for maintaining foreign R&D units and the *historical processes* through which such units are created (Figure 5.1).

---

[3] In principle, the survey encompassed R&D performed in majority-owned foreign subsidiaries. However, it includes the activities of Saab-Scania in Finland and of Volvo in the Netherlands, although these are not majority owned.
[4] For further details, see [12, 13].
[5] Of course, the exact definition of what activities qualify as 'R&D'—and the degree to which such activities are distinguished in accounting and reporting systems—vary considerably between companies and industries. Although there was no way to overcome this and similar problems of reliability common to studies of this kind, it is reasonable to assume that the quality of the database meets fairly high standards of accuracy.

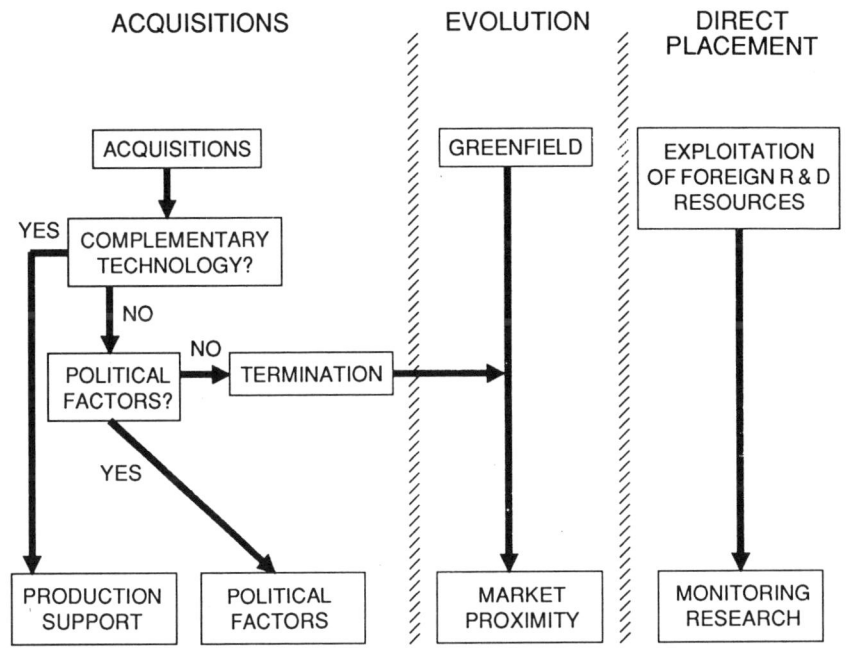

Figure 5.1 Internationalization of R&D

In the empirical analysis, five groups of foreign R&D units were identified, labeled according to the dominant motive for their establishment: (1) political factors, (2) production support, (3) market proximity, (4) monitor research and (5) multimotive units. With the exception of the last group, each type of laboratory tends to be associated with a specific type of establishment process: (1) acquisitions, (2) evolution of R&D activities in greenfield subsidiaries, and (3) 'direct placement' of research units.

5.3.1 Acquisitions

A very large proportion of foreign R&D units are found within acquired companies, reflecting the widespread and increasing reliance on external growth as a method of international expansion [12, 13][6]. Although, in theory, companies have the option of expanding or closing down activities in acquired units, reflecting, *inter alia*, the suitability of the particular location, such decisions and adaptations can generally be implemented only with a considerable time lag.

---

[6] During the period 1979–86, acquisitions accounted for 71% of all new foreign subsidiaries in Swedish industry [30].

Horizontal acquisitions. In the case of horizontal acquisitions, the work and competence of acquired R&D laboratories tend to duplicate that of already existing units. Economic deliberations may therefore suggest closing down acquired R&D units. Frequently, however, political and organizational considerations—including wishes and demand from local authorities and trade unions, legislation regarding job security, the wish not to upset employees in a newly acquired company, etc.—may make this impossible or unattractive. Hence, *political factors*—both internal to the company and reflecting government influenced conditions in the local environment—may significantly influence the decision to retain R&D in acquired companies.

Thus, it may be necessary or advantageous to be able to demonstrate to local authorities—especially if these authorities are also customers—that the company performs some local R&D. Maintaining R&D in acquired units can also bring other types of more or less politically determined advantages. This may, during recent years, have been the case for Swedish MNCs having acquired subsidiaries within the European Community. It is not unlikely that some of these companies have elected to retain acquired R&D capacity, awaiting the creation of the European single market in 1993.

Nevertheless, also when management in the acquiring company avoids a drastic cutdown in on-going R&D activities, the long term effect may be the same. In order to efficiently utilize the newly acquired technical capacity, the work of acquired units needs to be redefined and integrated with that of the rest of the group. The tasks allocated to acquired laboratories become subordinated to the overall needs of the group, often with an emphasis on adaptation of mother company technology. In terms suggested by Casson [3], the unit's primary mission would move from *generic*—developing new product or process concepts—to *adaptive* R&D, i.e. modifying applications with a specific local environment in mind. Such changes usually entail a reduction in the level of technical sophistication and circumscribe the freedom to take new technical initiatives. In consequence of such reorganizations, some of the unit's best and most qualified technicians may decide to resign, seeking other, more promising jobs—a decision which in many countries is facilitated by favorable labor market conditions for skilled engineers. In this manner, acquired R&D laboratories may 'dissolve themselves' without any need for drastic parent company intervention.

To the extent that the efficiency of foreign R&D operations is influenced by locational characteristics, such processes will lead to a more 'optimal' or 'rational' location pattern—but only with some, perhaps considerable, delay.

Product diversification. The situation is different when foreign acquisitions are undertaken as part of a strategy of diversification. In such acquisitions, the technical capacity and competence of acquired R&D often constitute their very *raison d'être*. As a rule, parent company efforts will then be devoted to retaining and perhaps expanding acquired technical manpower and resources. The primary task of such *production support* units is to perform generic R&D related to products and technologies unique to the company.

5.3.2 Evolution

With growth and age, successful foreign subsidiaries accumulate financial and technological resources, as well as 'political clout' within the corporation. This may lead to the establishment and growth of local R&D capacity. Local subsidiary management often has incentives to upgrade routine technical activities into product and process development. Local R&D may be a means both to create new business opportunities and to facilitate recruitment of qualified technical expertise. Impulses for such R&D emanate in the commercial relations with the local market: specific customer needs that cannot immediately be solved, unexpected results from technical activities that can be commercially exploited, etc. As a result, the dominating rationale for this type of unit is *market proximity*.

Typically, Swedish multinationals sell technically sophisticated goods on industrial markets, often requiring a high degree of customization and technical service. Foreign subsidiaries, therefore, tend to possess high engineering competence, making the transition from technical support into design and development—and from adaptive to generic R&D—almost inevitable[7]. In Swedish MNCs, this process was —at least historically— facilitated by the high autonomy traditionally given to local subsidiary managers, and by rather loose and informal control systems [8]. In many cases, local R&D capacity was created without formal authorization from headquarters[8].

The probability for the occurrence of such 'spontaneous' evolutionary processes is clearly in part determined by locational characteristics, such as the size, growth and sophistication of the local market. However, the evolution of local R&D in a foreign subsidiary also depends on 'random' factors, such as the entrepreneurial talent of local subsidiary management and 'luck' in the initial selection of development projects.

---

[7] As noted by Steele [29, p. 213]: "Once some technical capability has been established, domestic scientists and engineers are certain to see additional opportunities for improvement. There is an almost irresistible creepage from production engineering upstream into design and development". The significance of this type of evolution was emphasized also by Ronstadt [26, 27].
[8] This is in contrast to the situation in most US multinationals, where foreign R&D units are typically only set up after careful evaluation and scrutiny from corporate headquarters [26, 27].

5.3.3 Direct Placement

Sometimes, albeit rarely, the decision to set up a foreign R&D establishment is preceded by strategic type deliberations. This appears to be especially common in the case of units set up to perform long range, generic research. Since their work often has no immediate relevance to on-going operations, such units tend to maintain only very tenuous linkages to other corporate functions. In selecting a location for such units, the locations of existing plants and operations are largely irrelevant[9]. Instead, focus is placed on the possibilities to *monitor research* and 'tap into' a foreign scientific infrastructure, especially in terms of recruitment (large labor market pool of technical experts, amenable living conditions, etc.) and the possibility of maintaining close contact with other scientific organizations.

Since the establishment of long range research centers is usually seen as a strategic investment decision, it is reasonable to assume that a great deal of attention is given to the various factors that may affect the efficiency and effectiveness of the unit. Contrary to most other types of foreign R&D units, explicit consideration of locational alternatives can be expected, and final locations should reflect executives' notions of the locational characteristics likely to be beneficial to the performance of long range research, proximity to leading universities and other significant scientific institutions, availability and cost of skilled scientists and engineers, etc.[10]

5.3.4 'Multi-motive Units'

In the analysis, a rather large proportion of foreign R&D units could not be allocated to any of the aforementioned groups, suggesting that—similar to home-based R&D laboratories—they had been set up or retained for a combination of considerations and motives. This interpretation is supported by the fact that the units in this group seem to perform a substantial amount of advanced and innovative R&D. A fairly high proportion of their personnel holds postgraduate degrees and a relatively high share of their resources are devoted to research. Almost half report—directly or in an implied matrix structure—to headquarters staff in Sweden, again an indication that their tasks involve generic research of corporate significance.

---

[9] Indeed, it is sometimes argued that such units *should* be located away from production plants and other operative units in order to prevent them from becoming too involved in day-to-day technical problem solving. For units having a corporate mission, i.e. long range research for the group as a whole, a peripheral location may also help to emphasize their 'neutral' status *vis-à-vis* divisions and major subsidiaries.

[10] Especially in the case of expatriate engineers, recruitment possibilities are closely related to the level of net salaries prevailing. Hence, personal income tax structures are frequently an important aspect of the location, perhaps more so in Swedish companies than in multinationals with their home base in countries with less progressive income tax rates.

5.4 LOCATIONAL DETERMINANTS

R&D tasks are typically unstructured and involve negotiations, persuasion and joint problem solving, activities that tend to require frequent face-to-face contacts. Hence, *physical distance* and *travel time* to parties providing significant *informational inputs* to the development process can be expected to affect the attractiveness of potential R&D locations. Such inputs can be provided by a wide range of parties, both internal and external to the firm. Their role and consequences can be expected to vary over the life of R&D projects, but also with the primary mission of a specific R&D unit. Hence, in the case of market-driven R&D, proximity to important customers is likely to be important. However, as in most product development activities, significant inputs are probably provided also by the company's own marketing personnel, as well as by manufacturing engineers. In contrast, for long range research without immediate product applications, proximity to universities and other research establishments may be critical. Sometimes, work needs to be coordinated and adjusted to that going on in other R&D units, and proximity to other laboratories may be vital. Since most R&D activities have strategic significance, contacts with divisional or corporate top management are often important.

The cost and efficiency of an R&D laboratory will also be influenced by the price and availability of *material inputs*, including salaries for engineers and other personnel, cost of equipment and buildings, etc. However, international comparisons of R&D costs are not only difficult[11], but relative cost levels between countries change relatively rapidly in response to exchange rate movements, inflation rates and other factors. Thus, whereas availability of specific technical expertise and local labor market conditions are often significant, relative cost levels are probably a secondary concern to decision makers[12].

R&D locations may also be influenced by various *political factors*, i.e. factors directly or indirectly influenced by government authorities and policy makers. Although direct host government subsidies to R&D do not seem to be a major influence [9, 10, 22], other considerations—labor legislation, demands from local authorities and trade unions, etc.—may make it advantageous or necessary to establish some local R&D, or, more commonly, retain R&D in acquired companies. Indeed, in some industries, where governments or government-controlled organizations are important customers (e.g. telecommunications and military equipment), the assurance of local production and R&D capability may be a strong argument in negotiations and sometimes a prerequisite to being granted a contract.

---

[11] Frequently, official exchange rates do not accurately reflect real costs of R&D activities.
[12] Possibly with the exception of laboratories employing a high proportion of expatriates, for whom prevailing net salary levels are likely to be important (cf. footnote 10 above).

In general terms, geographical decentralization of R&D is a consequence of the historical growth and internationalization of company activities. Hence, locational patterns are also affected by *organizational* and *behavioral* factors [14], sometimes reflecting historical 'accidents', such as major acquisitions. However, some also have a predictable geographical dimension. Thus, companies may be more inclined to perform R&D in reasonably familiar types of environments, i.e. countries whose language, customs and legislation do not differ too much from those of the home country, thereby facilitating communication and information exchange.

Against this background, the influence of five different factors on the location of foreign R&D establishments in Swedish MNCs will be analyzed: *local market size, host country scientific infrastructure, need for market adaptation, political factors*, and *'psychic distance'*. The following sections outline the assumed influence of these factors (Table 5.1).

Table 5.1 Summary of hypotheses

DEPENDENT VARIABLES	INDEPENDENT VARIABLES Foreign R&D employment in:					
	All units	Market oriented units	Production support units	Research oriented units	Politically motivated units	Multi motive units
Local market size	+	+	+		+	+
Host country scientific infrastructure				+		
US location			+			
Location in one of the six original EEC countries					+	
Psychic distance from Sweden	−	−	−	−	−	−

5.4.1 Local Market Size

The size of the local market can be expected to influence not only the location and scale of foreign manufacturing, but also the performance of local R&D. In smaller subsidiaries on marginal markets, the volume of adaptive R&D is usually too small to warrant the establishment of local technical capacity, as required and economically feasible adaptations of products and processes will be performed by home country technical personnel. However, with growing size of local production, the volume and frequency of technical adaptations and new product development increase and it may become economically attractive to perform such activities in a local R&D unit.

Moreover, on average, the size and economic importance, as well as the technical and financial capacity of foreign subsidiaries are likely to vary with the size of the local market. Large subsidiaries are not only more likely to perform local R&D than are small ones; their number of R&D employees is also likely to be higher.

As a first hypothesis, therefore, *local market size is expected to positively influence the location of all kinds of foreign R&D*. However, foreign laboratories set up to monitor research are an exception. The locations and activities of such units are often not linked to local production but are primarily influenced by the possibility of recruiting local technical expertise and of maintaining contacts with local research establishments. Their mission is often to improve the technical capacity of the corporation as a whole. The scope of potential applications is often worldwide, and the size of the local market therefore irrelevant.

5.4.2 Host Country Scientific Infrastructure

Favorable cost, quality and availability of local engineers and other technical personnel tend to positively influence the performance of foreign R&D, regardless of scope or mission. Similarly, a dynamic technological environment—with a high level of interaction between companies, universities and other research establishments—should, in general, be conducive to the creation of an 'innovative climate' in local R&D establishments and may facilitate recruitment of skilled engineers. However, most foreign R&D units tend to be relatively small and their locations are predominantly determined by other factors. Hence, as a second hypothesis, *the presence of an advanced scientific infrastructure is expected to influence primarily the location of research oriented units*.

5.4.3 Need for Market Adaptation

In general, the need for market adaptation can be expected to be most acute in markets that in significant respects differ from the home market, for which most products were originally developed. For many consumer products, this may be the case in culturally dissimilar countries and/or countries with very different income levels. However, most Swedish multinationals produce and market industrial goods, many of which tend to be used in similar ways and are bought by similar customers around the world. Necessary customization, although often requiring relatively advanced technical competence, can, as a rule, be performed by manufacturing and marketing staff and does not require the formal establishment of proper R&D.

Sometimes, local cost structures and/or availability of raw materials and components may require adaptations of manufacturing processes. Such adaptations may also be needed to scale down (or up) production to levels different from those of other production sites. However, many such process adaptations need only be performed once, i.e. when production is first set up in a particular location, and are often undertaken by home country R&D personnel.

Similarly, product adaptations to market characteristics significantly different from those of the Swedish home market, e.g. such as those prevailing in most third world countries, are often undertaken by parent company technicians, especially since activities in such markets are often too small to support a local R&D department.

For Swedish multinationals, the main exception to the above considerations is probably the US market, which in many significant respects—technical standards, buyer behavior, market structure, etc.—differ from those encountered in most European countries [34]. In the case of US subsidiaries, both geographic distance from Sweden and the volume of local operations can be expected to induce companies to perform market adaptive R&D locally. Thus, the third hypothesis is that a *US location will tend to favor market oriented foreign R&D*. The hypothesis is tentatively supported by previous analyses of the data, showing 'market proximity' to be a primary motive in no less than 54% of the R&D employment in the North American units of the sample [12, 13].

5.4.4 Political Factors

As noted above, the significance of political factors as an inducement for performing R&D in foreign subsidiaries is partly industry-specific. In our sample a large proportion of the 'politically motivated' foreign R&D units belong to the Ericsson group and are active in the telecommunications industry. Many of these were set up in conjunction with the delivery of public telephone exchanges, in response to the demands of local PTTs. Their classification into the 'politically motivated' category reflects the fact that most PTTs are government owned. In the bidding for major orders of telecommunications equipment, local content requirements and R&D support are frequently important. However, such requirements are only partially motivated by political considerations such as balance-of-payment and employment effects. The efficient installation of new telecommunication systems often requires extensive technical adaptations and technical support to identify and solve initial 'bugs'. Hence, there are also good commercial reasons for customers to require the establishment of local R&D. 'Market proximity' may therefore be an equally significant motivation in many of these units.

Nevertheless, other types of more purely political factors—trade barriers and other forms of protectionism, the possibility of participating in government sponsored research programs, etc.—may still be important. Thus, establishing and maintaining a foothold inside the Common Market has long been a strategic concern for many Swedish companies. Although the 1973 conclusion of the free trade agreement between Sweden and the EEC may have relaxed the need to maintain production facilities inside the Common Market, subsequent developments—as manifested by the planned formation of the European single market—probably reinforced the perceived importance of a strong presence inside the Community.

Of course, the significance of this factor may have partly disappeared with the Swedish government's sudden change of policy vis-à-vis the Common Market in the fall of 1990 and the ensuing Swedish application for membership. Nevertheless, it can be expected that several of the European R&D units in the sample owe, at least partially, their existence to this historical rationale. As a fourth hypothesis, *a location inside the original six EEC countries is expected to have influenced the establishment of politically motivated R&D units.*

5.4.5 Psychic Distance

Several studies of the internationalization pattern of large Swedish multinationals [2, 17, 19] have detected a clear sequential pattern in terms of market selection and entry forms. Thus, companies tended first to enter geographically and culturally neighboring countries—Denmark, Norway and Finland—before gradually extending operations to more distant markets. Establishment sequences were shown to be strongly correlated with measures of the 'psychic distance'—defined as 'the sum of factors preventing the flow of information from and to the market' [18, p. 24]. In spite of the advent of modern communication technologies and the resultant increase in information flows, Swedish companies venturing abroad more recently still appear largely to follow this pattern, dictated by the need for an incremental learning process in order to reduce the uncertainty inherent in foreign operations [21].

As a fifth hypothesis, *the propensity to perform R&D in foreign locations is expected to decrease with the 'psychic distance' from Sweden.* On the one hand, in order to facilitate information exchange, companies are expected to prefer to carry out R&D in culturally familiar environments in geographically neighboring countries; on the other hand, and probably more important, foreign subsidiaries in these markets are, on average, older. They will therefore have had more time to grow and accumulate the technical competence and financial capacity needed to carry out local R&D.

5.5 EMPIRICAL TESTS

5.5.1 Data and Variables

The empirical testing of the hypotheses outlined above is based on data on the 151 foreign R&D units for which complete questionnaire responses were received. On the basis of a factor analysis of the weightings (on a five-point scale) of 21 potential motives for establishing R&D in a foreign unit (or retaining such activity in an acquired company), four sets of motives were identified, i.e. *'market proximity'* (accounting for 32% of total foreign R&D employment in the sample), *'production support'* (5%), *'political factors'* (34%), and *'monitoring research'* (8%). By means of cluster analysis of the corresponding factor scores, R&D units were classified into five categories. Four of these were characterized by large scores on one of the factors, i.e. one set of motives seemed to have dominated over the others and the category labeled accordingly. In a fifth group, accounting for about one-third of the laboratories (and 22% of the foreign R&D employment), none of the four factors dominated over the others. It was assumed that the units in this *'multi-motive'* category had been established or kept for a *combination of considerations and motives*.

In the following, the geographical distribution of foreign R&D employment, in total (TOTR&D) and by each of the five categories (MARKET, PRODUC, RESEAR, POLITI, MULTIM), is analyzed. Data are available for 21 countries, in which at least one of the 20 companies in the sample reported the presence of a laboratory.

Since the bulk of the companies in our sample are active in producer goods industries, *local market size* is measured by the size of the country's total *manufacturing employment in 1982* (MANEMP) [32, Table 98]. In view of the large variation in country sizes, and since the importance of this factor is not likely to be linear, the logarithm of this number was used.

The size and importance of *host country scientific infrastructure* was approximated by *total R&D expenditures as a percentage of gross national product* in the early 1980s (R&DINT) [31, table 5.20]. Of course, one may argue that the attractiveness of a particular country as a location for e.g. long range research may be partially determined by the absolute volume of R&D performed there—rather than by its relative importance to the economy. However, such an absolute measure would correlate strongly with country size and therefore with manufacturing employment, another of the independent variables. Moreover, relative R&D intensity is likely to better capture the quality of local science and technology.

The *need for market adaptation in the US* and the assumed *significance of political factors in the EEC* are measured by simple dummy variables (US and EG6), assuming the value '1' in the respective location, and '0' otherwise.

As a measure of *'psychic distance'*, an index computed by Nordström [21] was used (INDEX). The index, with a range from 45 (Norway) to 7500 (Thailand), was constructed on the basis of a simple questionnaire administered to a sample of Swedish executives attending various management programs organized by the Swedish Institute of Management (IFL). Although rough and based on a limited sample of executives, the reliability of the index appears to be acceptable. Thus, the relative ranking obtained correlates very strongly with the ranking obtained by Hörnell *et al.*, which is based on a much more elaborate procedure using various objective measures [17]. The correlation matrix between the variables is given in Table 5.2.

Table 5.2 Correlation between dependent and independent variables

	(1)	(2)	(3)	(4)	(5)	(6)	(7)	(8)	(9)	(10)	(11)
TOTR&D (1)	1.000										
POLITI (2)	0.213	1.000									
PRODUC (3)	0.416	0.002	1.000								
MULTIM (4)	0.928	0.195	0.234	1.000							
MARKET (5)	0.800	-0.108	0.590	0.671	1.000						
RESEAR (6)	0.208	-0.133	0.269	0.087	0.225	1.000					
INDEX (7)	-0.245	-0.067	-0.407	-0.444	-0.244	-0.260	1.000				
MANEMP (8)	0.471	0.196	0.140	0.395	0.461	0.114	0.307	1.000			
R&D INT (9)	0.424	0.027	0.395	0.362	0.428	0.521	-0.481	0.446	1.000		
USA (10)	0.593	-0.197	0.434	0.430	0.784	0.287	-0.086	0.425	0.372	1.000	
EG6 (11)	0.230	0.110	0.525	0.025	0.337	-0.016	-0.293	-0.129	0.118	0.461	1.000

5.5.2 Regression Results

Hypotheses regarding the locational determinants for overall R&D employment were tested by means of linear OLS regressions. The resulting estimates are given by equation (1) in Table 5.3. Only one of the hypothesized variables, market size, as measured by manufacturing employment, appears to influence the total volume of foreign R&D in the 21 countries. However, the level of significance is modest as is the overall explanatory value of the model.

Deleting the two variables with lowest t-values (country R&D intensity and the dummy variable for a location in one of the original Common Market countries) improves the model only marginally (equation 2). However, in this formulation, US locations appear to positively and significantly influence the performance of local R&D, as does proximity to Sweden in terms of psychic distance.

Testing the hypotheses regarding the determinants of different kinds of foreign R&D laboratories is complicated by the fact that the various types

Table 5.3 Regression results

Equation nr	(1)	(2)
Dependent variable	TOTR&D	TOTR&D
Constant	−606.724	−549.345
	(−1.024)	(−1.095)
MANEMP	185.817*	146.150*
	(1.784)	(2.006)
R&DINT	−111.853	
	(−0.597)	
US (Dummy)	843.908	829.359*
	(1.627)	(2.011)
EG6 (Dummy)	2.211	
	(0.009)	
INDEX	−0.087	−0.065*
	(−1.629)	(−1.824)
R^2	0.358	0.420
F	3.229**	5.826***
Number of observations	21	21

Figures in brackets are t-values.
Level of significance, two-tailed test: *** 1%, ** 5%, * 10%.

of R&D units are not present in all of the 21 countries; in all categories of foreign R&D, country employment is absent in several instances. Since the method of least squares can lead to potentially seriously biased estimates when applied to censored dependent variables, the hypotheses have been tested by means of maximum-likelihood estimation of standard TOBIT models [20]. The results are summarized in Table 5.4.

As hypothesized, market size is positively related to the volume of all kinds of foreign R&D, except laboratories set up to monitor research. Except in the case of multi-motive and research units, the relationships are statistically significant.

The presence of an advanced scientific infrastructure, as measured by country R&D intensity, is positively (and significantly) related only to R&D employment in research oriented units, thus supporting the second hypothesis. Somewhat surprisingly, local R&D intensity appears to be *negatively* related to the performance of politically motivated R&D, possibly because several of these units are located in third world countries. As expected, a location in the US significantly increases the probability of market oriented foreign R&D employment, but not that in any of the other categories.

Similarly, locations in one of the original six EEC countries significantly influence the performance of politically motivated R&D, supporting the assumption that Swedish companies have actively attempted to maintain and strengthen their technical capacity inside the Common Market. Although this finding sheds some light on the original motivations to establish or, in the case of acquired companies, retain R&D in these countries, it obviously

Foreign R&D in Swedish Multinationals ——————————— 113

Table 5.4 Tobit analysis: results

Equation nr	(3)	(4)	(5)	(6)	(7)
Dependent variable	MARKET	PRODUC	RESEAR	POLITI	MULTIM
Constant	−325.001*	−150.012*	−62.300	−169.746***	−22.603
	(−1.777)	(−1.752)	(−0.366)	(−2.654)	(−0.191)
MANEMP	78.642**	29.994*	−29.908	35.411***	24.196
	(2.335)	(1.818)	(−1.035)	(2.998)	(1.185)
R&DINT	−55.875	−5.421	164.285**	−31.948*	−22.343
	(−1.173)	(−0.203)	(2.253)	(−1.825)	(−0.580)
US (Dummy)	302.161**	−31.022	315.390	−263.590	153.437
	(2.292)	(−0.536)	(0.032)	(−0.029)	(1.515)
EG6 (Dummy)	−7.607	49.765*	−287.809	53.414**	−46.562
	(−0.111)	(1.742)	(−0.029)	(2.418)	(−0.941)
INDEX	−0.043**	−0.027**	−0.010	−0.102*	−0.0244**
	(−2.481)	(−2.404)	(−0.538)	(−1.893)	(−2.227)
SIGMA	83.912***	35.103***	62.501***	28.674***	70.923***
	(4.826)	(3.959)	(3.198)	(5.127)	(5.685)
Log-Likelihood	−80.608	−49.257	−35.995	−69.882	−98.967
Number of observations	21	21	21	21	21

Figures in brackets are t-values.
Level of significance, two-tailed test: *** 1%, ** 5%, * 10%.

gives no indication about the primary mission and tasks of these units. It seems probable that most focus on adaptive R&D, similar to that undertaken in market oriented units. However, it cannot be precluded that others also perform generic types of R&D, such as that undertaken in production support and multi-motive units.

Ease of communication, as indicated by 'psychic distance' from Sweden, appears to be a significant determinant for the performance of foreign R&D. The only exception concerns research oriented units. As suggested above, the activities of such units are often only very weakly linked to those of other corporate units. Information exchange with headquarters and other subsidiaries is not extensive, and other types of locational considerations—proximity to foreign research establishments, recruitment possibilities, etc.—override the importance of easy intra-group communication.

5.6 SUMMARY AND CONCLUSIONS

In summary, the statistical tests strongly support the notion that locational determinants differ between different kinds of foreign R&D activities. However (as expected), the variables included in the models account only for a relatively limited portion of the observed locational distribution of foreign R&D employment, suggesting the significance of 'historical accident'—R&D unrelated acquisitions, the activities of unusually entrepreneurial subsidiary managers, etc.—and other essentially 'random' factors. Moreover, the importance of organizational and behavioral factors is underscored also by the consistent (except in the case of research oriented units) significance of 'psychic distance' from Sweden as a determinant for foreign R&D.

In terms of overall foreign R&D employment, 'demand-related' factors —market potential and market adaptation in the US—are clearly more important than 'supply-related' factors associated with high host country R&D intensity. The latter seem to influence only the establishment of research oriented units. Of course, the economic significance of these units may be greater than their share of total R&D employment (8%). However, for the vast majority of foreign laboratories, other factors are clearly more important.

Finally, the apparent attraction of the six original EEC countries for 'politically motivated' R&D units lends some support to the notion that, in the past, Swedish MNCs have been led to perform a larger share of their foreign R&D in these countries than can be explained by the size of their respective markets and their relative proximity to Sweden. However, due to the somewhat dubious validity of the statistical procedure used for the classification and identification of 'politically motivated' foreign R&D units, this conclusion is, at best, tentative.

5.7 REFERENCES

[1] J. Cantwell Chapter 4.
[2] S. Carlson (1974), *Investment in Knowledge and Cost of Information*, Acta Academiae Regiae Scientiarum Upsaliensis, Uppsala.
[3] M. Casson (1990), Global Corporate R&D Strategy: A Systems View. Paper presented at the Conference on Technology and International Business, Stockholm School of Economics.
[4] A.J. Cordell (1971), *The Multinational Firm, Foreign Direct Investment and Canadian Science Policy*, Science Council of Canada, Special Study.
[5] A.J. Cordell (1973), Innovation, the Multinational Corporation: Some Implications for National Science Policy, *Long Range Planning* 6, 22–29.
[6] J.H. Dunning Chapter 2.
[7] O. Granstrand, C. Oskarsson, N. Sjöberg and S. Sjölander (1990), Business Strategies for New Technologies. Paper presented at the conference on 'Technology and Investment', in Stockholm, January, arranged by the Royal Swedish Academy of Engineering Sciences (IVA) in cooperation with OECD and The Swedish Ministry of Industry. Published in E. Deiaco et al. (eds), *Technology and Investment. Crucial Issues for the 1990s*. Pinter Publishers, London, pp. 64–92.
[8] G. Hedlund and P. Åman (1974), *Managing Relationships With Foreign Subsidiaries*, Sveriges Mekanförbund, Stockholm.
[9] G. Hewitt (1983), Research and Development Performed in Canada by American Manufacturing Multinationals, in A.M. Rugman (ed.), *Multinationals and Technology Transfer—The Canadian Experience*, Praeger, New York.
[10] J.D. Howe and D.G. McFetridge (1976), The Determinants of R&D Expenditures, *Canadian Journal of Economics* 9, 57–71.

[11] L. Håkanson (1981), Organization and Evolution of Foreign R&D in Swedish Multinationals, *Geografiska annaler*, Ser. B **63**, 47–56.
[12] L. Håkanson (1989), Forskning och utveckling i utlandet. En studie av svenska multinationella företag, *IVA-PM* 1989:1.
[13] L. Håkanson and R. Nobel (in press), Foreign Research and Development in Swedish Multinationals, *Research Policy*.
[14] L. Håkanson and R. Nobel (in press), Determinants of Foreign R&D in Swedish Multinationals, *Research Policy*.
[15] L. Håkanson and U. Zander (1986), *Managing International Research & Development*, Sveriges Mekanförbund, Stockholm.
[16] L. Håkanson and U. Zander (1988), International Management of R&D: The Swedish Experience, *R&D Management* **18**, 217–226.
[17] E. Hörnell, J.-E. Vahlne and F. Wiedersheim-Paul (1973), *Export och utlandsetableringar*, Almkvist & Wiksell, Stockholm.
[18] J. Johanson and J.-E. Vahlne (1977), The Internationalization Process of the Firm—A Model of Knowledge Development and Increasing Foreign Market Commitments, *Journal of International Business Studies* **8**, 23–32.
[19] J. Johanson and F. Wiedersheim-Paul (1975), The Internationalization of the Firm—Four Swedish Cases, *Journal of Management Studies* **12**, 305–322.
[20] G.S. Maddala (1984), *Limited and Qualitative Dependent Variables in Economics*, Cambridge University Press, Cambridge.
[21] K.A. Nordström (1991), *The Internationalization Process of the Firm—Searching for New Patterns and Explanations*, IIB, Stockholm.
[22] D.A. Ondrack (1983), Responses to Government Industrial Research Policy: A Comparison of Foreign-Owned and Canadian-Owned Firms, in W. Goldberg (ed.), *Governments and Multinationals*, Oelgeschlager, Gunn & Hain, Cambridge, Mass.
[23] P. Patel and K. Pavitt (1991), Large Firms in Western Europe's Technological Competitiveness, in L.-G. Mattsson and B. Stymne (eds), *Corporate and Industry Strategies for Europe*, Elsevier, Amsterdam.
[24] P. Patel and K. Pavitt Chapter 3.
[25] R.D. Pearce (1989), *The Internationalization of Research and Development by Multinational Enterprises*, Macmillan, London.
[26] R.C. Ronstadt (1976), International R&D: The Establishment and Evolution of Research and Development Abroad by Seven U.S. Multinationals, *Journal of International Business Studies* **9**, 7–24.
[27] R.C. Ronstadt (1977), *Research and Development Abroad by U.S. Multinationals*, Praeger, New York.
[28] S. Sjölander (in press), Strategic Technology Issues in Large Multi-Technology Corporations in Japan, USA and Sweden, *Journal of Technology Management*.
[29] L.W. Steele (1975), *Innovation in Big Business*, Elsevier, New York.
[30] B. Swedenborg, G. Johansson-Grahn and M. Kinnwall (1988), *Den svenska industrins utlandsinvesteringar 1960–1986*, Industriens utredningsinstitut, Stockholm.
[31] *UNESCO Statistical Yearbook* (1987), Unesco, Nancy.
[32] *UN Statistical Yearbook 1985/86* (1988), United Nations, New York.
[33] M. Wortmann (1990), Multinationals and the Internationalization of R&D: New Developments in German Companies, *Research Policy* **19**, 175–183.
[34] L. Ågren (1990), *Swedish Direct Investment in the U.S.*, IIB, Stockholm.

Chapter 6

Business Culture and International Technology: Research Managers' Perceptions of Recent Changes in Corporate R&D

MARK CASSON, ROBERT D. PEARCE
AND SATWINDER SINGH

The chapter examines recent trends in the organisation of global R&D using information derived from 27 interviews and from responses to a postal questionnaire. Following the weakening of the strategic position of central laboratories within large firms, there has been a growing tendency for the networking of research between major laboratories within large firms. It seems that the global research strategies of large US and Japanese firms are gradually converging, although Japanese research still remains more centralised than in Western firms. An important impetus to networking has been the need to rationalise research laboratories entering the firm through acquisition. In particular, European firms investing in the US have found it useful to concentrate product development in the US whilst leaving basic research focused on Europe. This is because the US represents the largest segment of the world market in most high technology industries. The main limitation to concentrating more applied research in the US is the combination of the high salary costs of US scientists and engineers and an unhelpful regulatory environment.

Technology Management and International Business: Internationalization of R&D and Technology.
Edited by O. Granstrand, L. Håkanson and S. Sjölander. © 1992 John Wiley & Sons Ltd

6.1 INTRODUCTION

This chapter examines recent trends in R&D from the perspective of both the parent and the subsidiary. It is based on a combination of questionnaire and interview evidence, collected as part of the project described in Sections 6.4 and 6.5. This chapter considers only the responses to open-ended questions —the structured responses are analysed in Section 6.4. These open-ended questions concerned recent and anticipated changes in the direction and organisation of research within the firm. Only a minority of respondents bothered to answer the open-ended questions, and some of the responses were clearly designed to persuade rather than inform. But the fact that the respondents took the trouble to write—occasionally at length—about these issues indicates that they were perceived as crucial by at least a subset of firms. Because of the modest number of responses, however, it cannot be claimed that the results are representative.

Most of the responses fall into clearly defined categories, and these categories form the basis for the tabulations below. The results are reported separately for headquarters and overseas laboratories. They are also classified by laboratory location—the four-way classification includes UK, US, Europe (including Scandinavia, Switzerland and Austria) and elsewhere (principally Japan, Australia and Brazil). The results for headquarters and subsidiaries are broadly consistent. But whereas headquarters respondents tended to concentrate on questions of changing corporate philosophy, the subsidiaries focused on frustrations about host government policies.

The interpretation of the written responses has been facilitated by 27 interviews with a subset of the respondents. Interviews were concentrated on firms which appeared to have relatively sophisticated strategies, provided intriguing written responses, or were statistical 'outliers'. Interviewees were promised anonymity. However, the structure of the interview sample, and a brief description of the companies involved, are given in the Appendix. Interviewees' opinions, though probably unrepresentative, were nevertheless extremely helpful in clarifying what lay behind some of the general concerns and anxieties expressed by other firms.

6.2 THE CHANGING SCOPE OF CORPORATE R&D

Tables 6.1 and 6.2 indicate a significant trend for corporate R&D to become more applied—there is less 'R' and more 'D', as many respondents put it. This trend is stronger in some cases than in others.

Headquarters laboratories in the US have undergone major changes in this respect. Post-war, some of these laboratories became major institutions for fundamental research. In the 1980s their roles have often been redefined to encourage more active support of product development in the companies' operating divisions.

The same trend is apparent in the headquarters laboratories of UK and European firms. But in some other countries—notably Japan and Brazil—the movement has been the other way. There is now a greater emphasis on fundamental research and less on applied. Leading Japanese firms are now establishing corporate research institutes on the US model to complement the factory-based development work on which they have traditionally relied. There seems, therefore, to be a convergence of R&D practice: the fundamentally-oriented Western firms are becoming more applied, and the applied-oriented Japanese firms more fundamental. This is consistent with the view that Western firms are investing in Japan to gain access to Japanese *corporate* know-how, whilst Japanese firms are investing in the West to gain access to institutions of fundamental research [6]. Firms in newly industrialising countries such as Brazil are following a similar path to the Japanese, but have not yet reached the stage of establishing foreign subsidiaries in the West. Most of the 'upgrading' of work in these countries so far is concerned with the work of foreign-owned subsidiaries there.

So far as respondents to the questionnaire were concerned, there were no statistically significant differences between US and Japanese parent firms in the propensity to undertake basic or applied research, or to focus on product as opposed to process development (see Table 6.2). This contrasts with other studies [4, 5] which suggest that Japanese firms are more inclined than their US counterparts to emphasise 'D' rather than 'R', and to focus 'D' on process rather than product-related work. Since these studies are retrospective views of the 1970s and 1980s, comparison with the questionnaire results for 1989 suggests that the process of convergence may well now be complete. It is probable, though, that this is true of only the very largest firms, of the kind included in our study. The legacy of the past may still be significant in small and medium-sized parent firms.

While differences between parents may be limited, however, the traditional strength of the US in fundamental corporate research is reflected in the response of foreign subsidiaries. Although US overseas affiliates are becoming increasingly applications-oriented, like their headquarters, foreign subsidiaries in the US are not. In the 1980s European and Japanese investors attracted to the US by the strength of the science base appear to have reoriented some of their acquisitions to do more fundamental research for the group as a whole. Thus while the transition to applied research remains valid at the corporate level, within the corporation there is a continuing trend to locate the more fundamental aspects of research in the US. There is a trend for R&D to become more diversified—especially in laboratories located in the US and UK. This is somewhat surprising. Current thinking in corporate strategy suggests that firms would have become more

Table 6.1 Current trends in R&D reported by headquarters respondents

	Total	Location of laboratory			
		UK	US	Europe	Other
Scope of R&D					
To fundamental from applied	12	2	5	1	4
To applied from fundamental	20	5	13	2	0
To applied from technical support	3	0	3	0	0
To technical support from applied	2	0	1	1	0
More diversified	14	5	7	2	0
More focused	6	1	2	2	1
Closer to customer	12	3	7	2	0
Closer to producer	3	0	2	1	0
More emphasis on quality	0	0	0	0	0
More 'full length'	1	0	0	0	1
External relations					
More university links	2	0	1	1	0
Fewer university links	0	0	0	0	0
More use of external contractors	0	0	0	0	0
Expansion of external contract work	1	0	1	0	0
More 'turnkey' work	1	0	0	1	0
Nature of work					
More multi-disciplinary projects	3	1	1	1	0
New techniques in R&D itself	3	1	2	0	0
Shorter development cycles	2	0	2	0	0

focused instead. But only a small number of respondents indicated that such was the case.

Part of the explanation lies in the need to redeploy researchers in mature industries such as steel and heavy chemicals. Another factor is the need to integrate backwards into new sources of supply—e.g. to investigate the use of wood in place of oil within the energy sector.

Another part of the explanation lies in the difference between growing and declining firms. Growth-oriented firms are concerned to explore the full commercial potential of their core technology. They have a new method or technique which is in search of problems to solve, and this leads them to a diversified programme of research. Research becomes diversified because they are following the logic of the 'scope economies' inherent in general scientific knowledge. Whilst they are focused in terms of scientific thought, therefore, they are diversified in terms of applications.

Declining firms, on the other hand, have typically abandoned fundamental research as too costly, too high risk and too long term. They lack secure funding and cannot cope with the difficult managerial judgements involved. They are 'sticking to the knitting' in terms of a narrow range of applications for which they know a market already exists. It seems to be the

Business Culture and International Technology

Table 6.2 Current trends in R&D reported by foreign subsidiary respondents

	Total	Location of laboratory			
		UK	US	Europe	Other
Scope of R&D					
To fundamental from applied	14	3	8	2	1
To applied from fundamental	22	8	8	4	2
To applied from technical support	2	1	0	1	0
To technical support from applied	2	1	0	1	0
More diversified	13	3	8	0	2
More focused	5	1	2	1	1
Closer to customer	10	5	4	1	0
Closer to producer	5	2	1	2	0
More emphasis on quality	4	1	2	1	0
More 'full length'	3	0	3	0	0
External relations					
More university links	3	0	2	1	0
Fewer university links	1	0	1	0	0
More use of external contractors	1	1	0	0	0
Expansion of external contract work	5	1	3	0	1
More 'turnkey' work	0	0	0	0	0
Nature of work					
More multi-disciplinary projects	0	0	0	0	0
New techniques in R&D itself	7	2	3	2	0
Shorter development cycles	2	1	1	0	0

minority of declining firms that report that they have become more focused. (One reason that they may appear as only a small minority is that responses are somewhat biased towards the more successful firms.)

Another characteristic of the successful firms with diversified R&D seems to be a trend towards strengthening university links. This is hardly surprising, given the more fundamental nature of their research and the difficulty of managing a diversified programme without external assistance. Several Japanese firms have recently set up basic research laboratories in UK university science parks, for example. One UK-based laboratory reported increased use of external contractors generally.

It is interesting to note, though, that few managers referred to universities as a source of imaginative new ideas. They were perceived rather as a source of highly specific technical competencies, which could be used to break bottlenecks or overcome particular stumbling blocks. This applied not just to basic research, moreover, but often to process research as well. There is evidence that many corporate R&D managers preferred to do long range speculative thinking for themselves.

A characteristic of declining firms is the expansion of work as an external contractor. Whilst several successful firms indicated that they achieved fuller utilisation of expensive facilities (such as engine test-beds) by renting them out to other firms, the future of these laboratories clearly did not rest on selling their general scientific expertise on the open market. In an expanding firm there are plenty of internal demands for new product development to keep the research staff fully occupied. The renting out of equipment is simply a reflection of the high cost of operating (and depreciating) the most sophisticated equipment available. But in a declining firm the laboratory may well lose internal custom for product development and so be forced to sell its services outside if it is to survive. One manager in this position expressed the view that offering to consult for more sophisticated firms in related areas was a good way of getting ideas and finding out what potential rivals were up to. What benefit his customers derived from his company's advice is difficult to say!

Across the board there was a strong feeling that researchers needed to get 'closer to the customer'. In practice this seems to mean better liaison with the marketing department, greater participation in pre-market testing, and more involvement with the launch and post-launch warranty problems. When the firm has a small number of major customers it can also mean more involvement in customised design, and closer liaison with researchers and production managers in the customer institution.

There was much less enthusiasm for getting closer to the producer. Many of the R&D managers interviewed feared day-to-day involvement with 'firefighting' production problems. Only 5% or so of such incidents raised any serious issue about plant design, claimed one manager (the inference being that the usual cause was inadequate training of production workers). It was also suggested that close involvement with production encouraged the static and unimaginative attitude that the status quo in production technology was the natural state of affairs. In several cases the laboratory had recently moved to get away from the local production site. Other plants were often suspicious that the laboratory gave superior technical support to its local plant, and this soured relations within the firm. Removing the laboratory from a plant site could therefore help the laboratory to 'get closer' to other plants in the group.

Several laboratories reported that they had undergone considerable changes in methods of research. The main factor was advances in computing. These included the use of computer models in place of physical simulations, a related change from analogue to digital computing, and the coordination of researchers through networked work-stations accessing a common core of data. In the pharmaceutical industry genetic engineering techniques had become very important too.

Other trends reported include a greater emphasis on the quality of research (presumably a spin-off from current trends in general management thinking) and greater use of multi-disciplinary research teams. Some firms noted that they now took 'full length' responsibility for research projects, seeing things through from start (initial concept) to finish (post-innovation performance), rather than concentrating on just product development. Firms in the electronics industry reported a striking shortening of product development cycles from 18 months or more to a mere 12 months. This reflected an accelerating pace of innovation, with a high frequency of incremental product up-grading. Pharmaceutical firms, on the other hand, were concerned that if their development cycles grew any longer there would be insufficient time to appropriate monopoly rents before their patents expired.

6.3 TRENDS IN ORGANISATIONAL STRUCTURE

There is a continuing debate in the organisation of R & D between those who favour putting all R & D into the operating divisions to ensure its practical relevance, and those who favour a separate central facility which is detached from operational pressures and can take a long-term strategic view. This debate surfaced in the US chemical conglomerate Du Pont as early as 1902 [3].

The argument in favour of a central laboratory is strengthened if it can be linked to other central services which provide complementary strategic inputs on marketing and finance. In one reported case, the central laboratory itself was involved in business and economic forecasting, while in another the director of the laboratory advised the main board on acquisitions.

The disadvantage of the central laboratory is that it can be 'hijacked' by an academically-oriented scientific elite. The danger is mitigated in some laboratories by giving scientists up to 10% 'playtime' during which they can carry on their own private projects. Many central laboratories also encourage publication, though crucial measurements (such as the temperature of chemical processes) are usually withheld. The existence of a central laboratory also creates a problem of multiple allegiance in the division laboratories, whose heads need to report both to the head of the central laboratory and the head of their own division. Such problems can arise even in the absence of a central laboratory, but they can be more difficult to resolve if the senior R&D director is the head of a central laboratory which he regards as his 'home constituency'.

The 1980s was a difficult decade for managers who wished to defend the central laboratory's prerogative to conduct pure research. They had to rebuff both the threat of closure, and the threat of 'capture' by the divisions, who would re-deploy them to development work.

In practice, though, compromise solutions were possible. In many growing firms fundamental research was distributed amongst all the major laboratories in the group on the basis of merit. The central facility then became just another facility within an international network of laboratories of equal status. The only distinguishing feature of the central laboratory that remains is that it handles particularly difficult scientific problems referred to it by the other laboratories. Its staff are typically better qualified, and represent a greater diversity of professional specialisms than others.

In a very large firm only a proportion of the laboratories may be members of the network. The formation of a network is sometimes associated with the downgrading of certain laboratories to carry out only local production or marketing support. This top-down 'rationalisation' of research was quite prominent in the 1960s and 1970s, but seems to have become less common recently, though. Modern research methods frequently allow subproblems —such as the writing of special software—to be subcontracted to almost any laboratory where capable scientists are employed. Thus research laboratories in developing countries can still do important work even though they may be short of sophisticated hardware. Laboratories in India and Brazil, which employ highly educated scientists, need no longer be peripheral to global research. The trend seems to be for networks to be extended to include many more of the laboratories within a group.

In several companies there were two dominant laboratories—one in Europe and one in the US. The most interesting case is where the US laboratory has been recently acquired by a European firm. There is anecdotal evidence that some European managers have difficulty influencing their US counterparts —who behave as if they were superior to, rather than formally subordinate to, the European management.

A more extreme form of network is to constitute the central laboratory as a profit centre rather than a cost centre, and relegate it formally to a consulting role. The laboratory receives only a proportion of funds from the group as a whole—say 25%—and relies on 'internal sponsorship' for the rest. This 'customer-contractor' principle requires the laboratory manager to go out and sell his services within the group. In a declining group, internal sales may be unsuccessful, and the laboratory may then move to external contracts to maintain its viability (see earlier). In some instances the development of a large external customer base has led to consideration of a 'management buy-out' by the scientific staff.

The network principle may also extend to laboratories that are not wholly owned by the group. The networking of joint venture partnerships has become quite fashionable in the 1980s [2]. Joint ventures allow a firm with limited capital and managerial capacity to extend the scope of its influence very markedly provided it is willing to sacrifice formal control. When the individual partners are much smaller than the core firm then loss of control need cause little concern, however. The joint venture approach is sufficiently flexible that it can be extended to include non-profit research associations as well as ordinary private firms.

Business Culture and International Technology ——————— 125

Table 6.3 indicates that while networking is very much the dominant trend, many headquarters managers retain the view that some degree of hierarchy between central and divisional laboratories is essential. The reorganisation of the central laboratory as a consultancy involves some firms in the UK and Europe but very few, it would seem, in the US. The evolution of headquarters into a nexus of joint venture partnerships is even more of a minority trend, it would seem. Our sample is, however, too small to draw any firm conclusions. But interview evidence suggests that inter-firm collaboration in R&D is still viewed with suspicion in many firms.

The foreign subsidiaries' perspective on organisational change is shown in Table 6.4. The trend towards networking is confirmed by the strong movement towards greater interdependence within group research. The interdependence applies to interactions with both headquarters and other subsidiary laboratories. The trend towards more formal mandating of subsidiaries for global or regional product development is more limited. One Canadian subsidiary of a US firm specifically complained that mandating was inflexible. Because of government pressures, products developed in Canada had to be first produced in Canada, even though Canada was not a good production site. The linking of research to production in government-imposed mandating may actually work against the interests of a laboratory, it was claimed. Research that the laboratory is well qualified to do may be done elsewhere because the parent firm does not wish to be tied to subsequent production in the country concerned.

A few managers said they expected their laboratories to achieve greater independence. These laboratories seemed to belong to firms where hierarchical coordination was still in use. Independence was seen as a way of achieving self-sufficiency in fundamental research, so that headquarters would simply leave them to get on with their projects alone.

6.4 THE PRACTICE OF R&D MANAGEMENT—EVIDENCE FROM INTERVIEWS

R&D strategies seem to be in a continual state of flux. Several respondents referred to cycles in which the emphasis of research shifted from pure to applied, from upstream problems of supply shortage to downstream problems of product innovation, and so on. Some reported that their laboratories had seen 'massive' changes—surviving only because of the flexibility of their employees.

An increasing impact of social concerns was evident. Tobacco firms were preoccupied with researching into the links between smoking and health, cement firms were investigating ways of extracting gravel with minimal environmental damage, food laboratories were working on the reduction

Table 6.3 Trends in organisation structure perceived by headquarters respondents

	Total	Location of laboratory			
		UK	US	Europe	Other
Maintenance of hierarchy	18	7	5	4	2
Development of network	24	8	10	5	1
Reorganisation as consultancy	6	2	0	3	1
Nexus of partnerships	2	0	1	1	0

Table 6.4 Trends in organisation structure perceived by subsidiary respondents

	Total	Location of laboratory			
		UK	US	Europe	Other
Greater independence with other labs					
Without special mandate	16	1	10	1	4
With special mandate	7	4	2	1	0
Greater independence	4	2	1	1	0
Greater subordination	0	0	0	0	0

of fat content and the elimination of additives, whilst information technology specialists were researching the impact of computer networks on the social organisation of groups. Other problems that engaged researchers included combating food poisoning in supermarkets, preventing sabotage in chemical plants and developing protocols for dealing with HIV infection amongst production line workers in the food and pharmaceutical industries. In some cases the research on these issues was perceived negatively —as an additional cost and as a distraction from more interesting lines of work. In other cases they were seen as providing profitable opportunities for the innovation of safer or 'friendlier' products.

A number of laboratories had been bought and sold several times, as a result of mergers, acquisitions, divestments and buy-outs. One manager indicated that his laboratory had been sold four times since he had worked there, and that it was currently up for sale again. Subsequently, it was bought and then resold again within a month.

The immediate cause of many of these changes was the financial position of the company. A minority of managers felt reasonably secure. Hitachi R&D in the US, for example, is governed by the philosophy that 'though we cannot live one hundred years, we should be concerned about one thousand years hence'. At least, that is the company's official line, and it is one that is corroborated by the evidence on the stability of Japanese R&D expenditure growth reported in Casson [1, Chapter 1]. Managers of US- and UK-owned

Business Culture and International Technology ——————— 127

laboratories do not seem so assured. 'R&D is the first to be cut back—it's a soft target' was a common complaint. Cutbacks were expected not only in times of exigency, however, but whenever a bid for the company was in the offing. Laboratory managers clearly believe that 'short-termism' in equity markets makes general management nervous about 'over-investing' in prestigious long-term R&D.

In the pharmaceutical industry funding can be particularly volatile. Even large companies often derive a substantial proportion of their profits from a single product—and the profit flow can dry up completely once the patent expires. The funding of R&D as a fixed percentage of sales, noted in Casson [1, Chapter 7], can only increase anxieties in such a case. Following the Beecham–Smith Kline merger, a further wave of mergers was confidently predicted—not all of which, it was suggested, would succeed. A shake-out of the less successful laboratories seems to be anticipated in the industry.

Notwithstanding these difficulties, morale in many of the laboratories we visited seemed quite high. Some subsidiary managers were keen to explain the devices they used to hide away reserves. New equipment was ordered in advance of requirements when profits were high, and budget restrictions were consequently lax, and replacements deferred when times were hard. Most R&D managers, like their counterparts in general management, had invested considerable effort in learning how to 'play the system'.

Short-termism is not confined to capital markets, according to our interviewees. Marketing executives too have short time horizons, and these are difficult to reconcile with the planning cycle in R&D. Because they lack scientific training, marketing executives are inclined to ask for the impossible—and to want delivery by the end of the week. One company has attempted to solve this problem by a promotional spiral, in which managers alternate between posts in marketing and research. Managers following one another up this spiral normally face an opposite number who has had experience doing their sort of job.

A consequence of putting researchers onto a general management track, however, is that they are removed completely from the laboratory bench. The majority of managers interviewed were scientists who had begun their career at the bench (often with the same company) and had then moved onto the management track. It was fairly unusual to find a non-scientist managing R&D. Some interviewees claimed that a non-scientist could not do their job at all because he could not win peer group respect.

Scientists who do not move onto the management track often face a 'dead end' so far as pay, prestige and power are concerned. There seems to be a widespread view amongst senior managers—at least in the UK—that scientists are motivated by the enjoyment of work and involvement in the local community, rather than by pay. They are seen as risk-averse individuals who like to remain close to the major agglomerations of laboratories, so that if they need to change jobs to advance their career they can do so without moving far. They are also believed to be relatively immobile, so far as overseas secondment is concerned, once they enter early middle age. According to this stereotype, mobility is characteristic only of junior scientific staff.

To improve the career prospects of the dedicated scientist, one UK firm had instituted a competitive scheme for high-profile research fellowships tenable at any of the company's laboratories. This can take respected research professionals outside the conventional hierarchy of pay and authority. Another company had turned down a similar scheme, and the manager concerned believed that some of his key scientists would leave to take up university professorships as a result.

Most of the R&D managers we interviewed clearly belonged to the 'human relations' school of personnel management. Researchers had to be trusted to get the best out of them, it was claimed, and anyone who could not be trusted should not be employed at all. One manager likened his job to that of managing a football team (he was a Liverpool supporter it should be noted). Most managers were also familiar with business strategy concepts such as first-mover advantage, niche marketing and fit. The only striking weakness was that few had a clear strategy regarding technology licensing. 'Never license your technology unless you have something better in the pipeline' was the only coherent strategy we were told about. The main impression was that licences are often *ad hoc* agreements negotiated at board level, with limited consultation with R&D. On the whole it would seem that licences, joint ventures and other formal kinds of external collaboration are 'top-down' initiatives which are disliked by practitioners of R&D.

Overall, though, the research managers we met were extremely self-confident individuals who seemed to combine the roles of technologist and 'intrapreneur' quite successfully. They claimed to have a good awareness of what research rival firms were engaged in. Indeed, they seemed to share information with their opposite numbers fairly freely, in terms of the kind of work going on. Whilst patent races were a recognised feature of certain industries, everyone seemed to have a good idea of who was in the race and whether it was going to be a tight finish or not.

Despite this strong competitive element, though, there was a widespread view that it was unethical to poach staff, either to disrupt a rival's project or to learn secrets from him. Indeed, it was claimed that new recruits would not expect to be systematically 'debriefed' on their former employer's operations. Conversely, people who were about to leave would not be ostracised or have their security clearance reduced. This is very different from the way that professional staff are handled in, for instance, international banking.

Underlying this ethical view was, apparently, a view that 'tit for tat' response was likely. A firm that head-hunted key personnel to disrupt a

Business Culture and International Technology — 129

rival's research would find its own staff poached in return. It is even possible that the leading firms might conspire to punish a deviant firm by excluding it from the informal information network. Objective evidence for the existence of a strong informal network comes from the coherence of collective lobbying over issues such as patent lifetime and government funding of basic R&D.

Research managers do not customarily enjoy a high profile so far as the financial press is concerned. Nevertheless, our interviews suggest that the decisions these managers make are often crucial for the long-run profitability of the firm. These decisions are subtle and complex, and often involve technical matters on which laymen cannot easily judge. But what can be judged is the way that the most successful managers combine their scientific knowledge with a grasp of financial issues and an intuitive feeling for personnel management. However competent the personnel manager, it is always possible for him to be marginalised because of prejudices against technology or long-term thinking at the highest levels of the firm. Firms which tend to marginalise R&D in this way will find it difficult to retain the most able staff, and so they will finish up with the quality of R&D management they deserve. There is anecdotal evidence from the interviews that the most influential R&D managers were also the most energetic and enthusiastic, but it is difficult to say exactly where cause and effect lie.

An important point that emerged from the comparison of the UK and overseas interviews was that it was more common to find the 'R&D culture' permeating the whole firm in the US and EC than in the UK. It was in North America that we met the President of the company rather than the R&D manager because the President considered himself the executive in charge of R&D. Senior management in his company was dominated by engineers, who sold the products to other engineers who managed procurement in the client companies. Other evidence suggests that this engineering culture is fairly common amongst US, German and Scandinavian parent firms. We encountered no instance of it in parents or subsidiaries in the UK.

6.5 GOVERNMENT POLICY

Headquarters respondents had relatively little to say about government policy—perhaps because their concerns were global rather than national, or perhaps because they were simply out of touch with the frustrations of managing a foreign-owned subsidiary in an alien environment. Subsidiary respondents, on the other hand, had quite a lot to say. Their views are summarised in Table 6.5.

The popular view of the US as a deregulated economy takes a sharp knock. Foreign investors complain about the amount of form-filling involved in

US operations, particularly in dealing with social matters. It is the Federal agencies which are most to blame. The Food and Drug Administration, for example, is considered unnecessarily stringent in its requirements for premarket testing. The cost of drug trials, it was claimed, can be up to ten times the cost of drug discovery and development. Bureaucratic demands absorb resources that could better be employed elsewhere. Occupational safety reporting requirements are also perceived as relatively stringent in the US. Industrial policy too is problematic in the US. Anti-trust measures are claimed to deter cooperative R&D with other firms. The commercialisation of spin-offs from government contracts is unduly inhibited, it is claimed;

Table 6.5 Changes in government policy desired by foreign subsidiary respondents

		Location of laboratory			
	Total	UK	US	Europe	Other
Reduced regulation					
General reduction	6	0	5	1	0
Faster decision-making	3	2	0	1	0
In social matters:					
relax reporting requirements in					
Product safety and pre-market testing	3	0	2	0	1
Occupational safety	2	0	2	0	0
Animal experiments	2	1	0	0	1
Environment	3	0	1	1	1
In industrial policy					
Fewer anti-trust restrictions on collaborative R&D	1	0	1	0	0
Less pressure for global mandating of subsidiaries	1	0	0	0	1
Less restriction on commercialisation of spin-offs from government funded R&D	1	0	1	0	0
Less restriction on export of key technologies	2	0	2	0	0
In marketing policy					
More flexibility over pricing	1	0	0	1	0
Abolish 'limited list' for prescriptions	2	2	0	0	0
Improvement of education and science base					
Increased support for university centres of excellence	3	2	0	0	1
More joint fellowships, studentships, etc.	3	2	0	0	1
More science graduates	2	2	0	0	0
Improve social status of engineering	1	1	0	0	0
Improve labour mobility through easier Immigration of research professionals	2	0	1	1	0

Source: Questionnaire.

enforced delays in patenting inventions, and restrictions on the intra-firm export of technology should be eliminated. Both Switzerland and the US are perceived as inflexible over the immigration of foreign research workers. The UK is believed by some to have a too-strict policy on animal experiments (which, like some of the other restrictions, is probably no bad thing from the social point of view). The fact that the UK is also, apparently, one of the few developed countries not to aggravate firms over environmental matters is a more dubious distinction, though.

Two pharmaceutical subsidiaries in the UK expressed concern about the 'limited list'—the reduction in the number of proprietary medicines which doctors are able to prescribe to National Health Service patients, and the consequent switch of government-funded demand to generic products. A related concern over pricing restrictions was expressed by a pharmaceutical subsidiary in Belgium.

Managers of subsidiaries expressed concern over the future of the education system in the UK, and the erosion of the science base. It is worth noting that no complaints of a similar nature were made in the US or elsewhere in Europe. Two firms called for increased support for university centres of excellence, and two for more joint research posts—and in particular the reinstatement of CASE (Collaborative Awards in Science and Engineering) scholarships. There was also a call for increased output of graduate scientists and an improvement in the social status of engineers (from a firm in the office equipment industry). It is also worth noting that in interviews, and at our preliminary conference, several research managers were emphatic on these points. Clearly the respondents were aware that they were telling academic researchers what they wanted to hear. But it is evident that they have been making the same points, quite independently, to government too.

6.6 GOVERNMENT FUNDING

Strong views were also expressed on funding issues, as indicated in Table 6.6. Unlike the previous responses, these were expressions of hopes and fears, rather than reports on actual developments. As might be expected, respondents were generally in favour of more government funding for R&D, though there were differences of opinion as to how this should be done.

There was a widespread view—particularly in continental Europe—that government should increase its funding for fundamental and high-risk research in the corporate sector. This would effectively subsidise long-term projects whose returns were difficult to quantify. The precise rationale for this view was not clear, however. It could reflect a belief that private returns from such projects are difficult to appropriate compared to the social benefits conferred (an externality argument) or that the government's obligation to future generations implied a lower rate of discount than management's obligations to its shareholders.

Some respondents sought more funding for corporate research of any kind, whilst others suggested that existing schemes to promote cooperative research (within the European community, for example) should be expanded. On the other hand, one UK respondent claimed that government discrimination in favour of cooperative research meant that his laboratory could no longer use government funding because of the threat to secrecy.

Several US-based laboratories favoured an expansion of Federal projects put out to competitive bidding. Another respondent complained, however, that such projects were always awarded to 'Beltway Bandits'—politically-networked firms that specialised in winning contracts but had no coherent research strategy of their own. This sceptical view was shared by several interviewees, who were adamant that government was a very poor judge of the kind of fundamental research that needed to be done. Instead of special projects, they favoured general incentives such as tax credits.

Tax credits on R&D expenditure seem very popular—they are a familiar fiscal instrument which subsidises R&D but leaves the corporation free to decide what research to do. Foreign subsidiaries in Australia seemed very impressed by the tight environmental restrictions imposed by some state legislatures. If the case for R&D subsidiaries is accepted by government, then tax credits seem to provide a simple and acceptable mechanism for implementing them.

Table 6.6 Fiscal changes desired by foreign subsidiary respondents

	Total	Location of laboratory			
		UK	US	Europe	Other
More grants					
for fundamental or highly innovative research	12	3	3	6	0
for private research generally	11	4	1	4	2
for cooperative research	6	3	2	0	1
More funding for specific projects					
defined by company	2	0	1	0	1
competitve bidding for projects defined by government	4	1	3	0	0
Tax incentives					
tax credits for R&D	10	2	5	0	3
lower corporate income tax	1	0	0	1	0
soft loans	1	0	0	1	0

Source: Questionnaire.

Business Culture and International Technology

6.7 APPENDIX

The 27 firms interviewed are cross-classified by ownership and location of laboratory in Table 6A.1. They are classified by industry in Table 6A.2. Each firm was identified by a number; a brief description of each firm is given below.

(1) A UK food company which has recently been bought and sold several times. It has been a subsidiary of British, US and European firms. It is oriented to producing confectionery and 'fast food' for the local market. R&D involves food science experiments conducted in large kitchens and process experiments conducted with a full-scale plant installed in the laboratory.

(2) A UK subsidiary of a UK conglomerate which concentrates on producing high-quality lubricating oils. Winning product acceptance in Germany is a key aspect of product development strategy.

(3) A UK tobacco company that carries out research to investigate links between smoking and health. Other activities include quality control, testing flavours and improving manufacturing process technology.

(4) A European cement company whose central R&D unit provides a consulting service for the group. It acts as a 'central intelligence agency' keeping constituent companies in touch with external developments.

(5) A UK cement company undertaking mainly applied research. The research concentrates on products rather than processes. The strategy is to be one of the leaders in technology but not necessarily *the* leader. The laboratory occasionally does external consulting work.

(6) A UK subsidiary of a US engineering company. R&D has a strong element of technical support for production. Operations are fairly small and innovation is mainly 'evolutionary' rather than 'revolutionary'.

(7) A UK food company that is stronger on marketing than on technology relative to the industry norm. The laboratory has switched from long-term to short-term projects, and has lost its key role as a prestigious research centre. It is sufficiently independent of the parent company that it could be easily divested or bought out by management.

(8) The UK subsidiary of a European pharmaceutical company. Like the other pharmaceutical companies, research is scientifically-driven and long-term. International coordination of R&D is informal but sophisticated.

(9) A UK textile firm that was one of the earliest UK firms to establish a large central laboratory. It engages in both basic and applied R&D.

(10) A UK drink company that has been actively involved in diversification through acquisition. The R&D director also undertakes technological audits of target companies.

(11) A UK subsidiary of a US oil company. It coordinates international R&D through relatively complex formal managerial procedures. The firm achieves high profitability through being a 'fast second' where innovation is concerned.

(12) The UK subsidiary of a US pharmaceutical company. It holds a global mandate for near-market product development, whilst basic research is carried out in the US.

(13) The US subsidiary of a Japanese chemical engineering company. Very little research is done in Japan. The main emphasis of R&D is on product development for a global market.

(14) A US glass company with significant diversification into high-technology fields. It has a well-established laboratory in Japan, and joint ventures involving collaborative R&D. Many scientific papers are published by the US laboratory.
(15) A US paper company whose laboratory undertakes long-term product development. It is highly geared to patenting.
(16) The US subsidiary of a diversified Japanese electronics firm. The laboratory is geared mainly to product development, with basic research being concentrated in other subsidiaries outside the US.
(17) The Canadian subsidiary of a US steel firm. During the 1980s the laboratory's shift from basic towards applied research was an unusually dramatic one. There is now strong interaction between the laboratory and the adjacent factory.
(18) The Canadian subsidiary of a US office equipment company. It has strong links with local universities and encourages multi-disciplinary research.
(19) The Canadian subsidiary of a US defence-engineering and telecommunications company serving major export markets.
(20) A UK pharmaceutical company which also has a major US laboratory. The two laboratories both carry out basic and applied research, but concentrate on different types of medical condition.
(21) The UK subsidiary of a US motor company. The laboratory shares responsibility for European product development with another laboratory in Europe.
(22) A European electrical engineering firm that has made substantial acquisitions in the US. The firm has collaborated with other European firms in product development. It was undergoing major organisational restructuring at the time of the interview.
(23) A UK oil company undertaking a well-diversified programme of research in the resources and exploration field.
(24) A European pharmaceutical company with a very strong patenting record and a reputation for 'openness' in management.
(25) A European drink company. Its R&D is mainly concerned with quality assurance, basic research being subcontracted to universities. Some research involves collaboration with rival firms.
(26) A European pharmaceutical company with a somewhat ethnocentric corporate culture.
(27) A European metals company, now extremely diversified. Research tends to be concentrated at the divisional level.

Table 6A.1 Ownership and location of sample laboratories

Location ownership	UK	US	Europe	Other	Total
UK	9	–	–	–	9
US	4	2	–	3	9
Europe	1	–	6	–	7
Other	–	2	–	–	2

Table 6A.2 Industrial composition of sample laboratories

Industry	Frequency
Food, drink and tobacco	5
Pharmaceuticals	5
Engineering	4
Paper, glass, cement	4
Oil	3
Electronics and office equipment	2
Metals	2
Textiles	1
Motor vehicles	1
Total	27

6.8 REFERENCES

[1] M.C. Casson (1991), (ed.) *Global Research Strategy and International Competitiveness*, Basil Blackwell, Oxford.
[2] P.J. Contractor and P. Lorange (eds) (1988), *Cooperative Strategies in International Business*, Lexington Books, D.C. Heath, Lexington Mass.
[3] D.A. Hounshell and J.K. Smith, Jr. (1988), *Science and Corporate Strategy: Du Pont, 1902-1980*, Cambridge University Press, Cambridge.
[4] E.S. Mansfield (1988), Industrial R&D in Japan and the United States: A Comparative Study, *American Economic Review*, **78**:2, 223–8.
[5] N. Roseberg and W.E. Steinmüller (1988), Why are Americans such Poor Imitators? *American Economics Review*, **78**:2, 229–234.
[6] D.E. Westney (1990), Internal and External Linkages in the MNC: the Case of R&D Subsidiaries in Japan, in C.A. Bartlett, Y. Doz and G. Hedlund (eds), *Managing the Global Firm*, 279–300.

_____ Chapter 7

Internationalisation of Research and Development among the World's Leading Enterprises: Survey Analysis of Organisation and Motivation

ROBERT D. PEARCE AND SATWINDER SINGH

This chapter presents survey evidence on the role of overseas R&D units in supporting the increasing use by leading enterprises of global approaches to innovation, which are themselves becoming more central to the overall global-strategies implemented by such firms. The results show that overseas R&D still has a widespread tendency to play its more traditional role of adapting a MNE's existing products, or production processes, to particular countries' markets or production environments. However, two possible new roles for overseas R&D were also investigated. Firstly there was substantial evidence supporting the view that overseas R&D laboratories often play a role in global programmes of innovation by deriving new products, which are then produced locally either for the host-country market or for wider markets, as part of World (or Regional) Product Mandate operations. Secondly, there was less clear evidence of the other possibility, namely that overseas laboratories of MNEs would use particularly high quality local scientific ability to play a role in internationally-integrated programmes of basic research. Nevertheless it does seem that where overseas laboratories do perform basic research it is then linked into wider programmes in the manner suggested.

Technology Management and International Business: Internationalization of R&D and Technology.
Edited by O. Granstrand, L. Håkanson and S. Sjölander. © 1992 John Wiley & Sons Ltd

7.1 INTRODUCTION

It is the purpose of this chapter to provide an overview of the results of a detailed survey investigation of the internationalisation of R&D by the world's leading enterprises[1]. The recent emergence of this area of research is predicated upon two beliefs. Firstly that overseas R&D now constitutes an important share of the total technological activity carried out by a significant number of the leading industrial enterprises. Secondly that the growth in such decentralised R&D operations reflects the perceived need by many MNEs for technology to play a role in direct support of an emerging global perspective on competitive strategy. This line of argument sees a globalised approach to innovation[2] as central to an effective global strategy, with dispersed R&D facilities as a crucial element in achieving this. The thinking which underlies this suggests that a traditionally accepted sequential approach, in which a centralised creation of technology was followed by diffusion and adaptation, has collapsed into a more internationally-integrated programme of research which recognises the need to respond continually (rather than through a technological 'trickle down' effect) to the needs of a range of important markets and also to involve geographically-dispersed heterogeneous scientific capabilities in fulfilling these targets. In this perception the decentralised R&D units may either contribute scientific inputs to the creation of new basic technology in the MNE group, or assist in the commercial implementation of such new knowledge by helping to facilitate a programme of more or less simultaneous differentiated innovations based on it, which acknowledges the idiosyncratic needs of different national markets.

Some broad indications of the geographical dispersion of creative activity in leading enterprises can be obtained through analysis of data on patenting in the USA[3]. The data on each patent records the location of the research facility originally responsible for the innovation, and the firm to which the patent has been granted. Once research[4] had established the ultimate ownership of patents, in cases where they had been granted to affiliates of MNEs, this meant that for each of a large number of these leading enterprises data was available on the total number of patents granted to the group in the

---

[1] For full presentation of the results see [9]. The survey was carried out as part of a research project funded by the Economic and Social Research Council.
[2] See [1].
[3] The data on the geographical origins of patents granted in the USA have been compiled at the University of Reading using data on patent counts obtained through the Science Policy Research Unit at the University of Sussex. The data on US patent counts were prepared by the Office of Technology Assessment and Forecast, US Patent and Trademark Office, with the support of the Science Indicators Unit, US National Science Foundation.
[4] Carried out in cooperation between researchers at the University of Reading and Science Policy Research Unit (University of Sussex).

Internationalisation of R&D: Organisation and Motivation

US in a particular year, and on the number of these that were attributed to research performed outside the MNE's home country. Thus we are able to derive the variable 'overseas subsidiary patent ratio' as 'the proportion of the total patents taken out in the USA in 1981/83 by a MNE which are attributed to the work of research units outside the MNE's home country'.

Of 694 enterprises for which overseas subsidiary patent ratio was calculated[5] it was found to be zero for 241 (34.7 per cent) and less than 10 per cent for 237 (34.1 per cent) more. However, 99 (14.3 per cent) of the enterprises had a ratio between 10 per cent and 30 per cent, 49 (7.1 per cent) between 30 per cent and 50 per cent, and 68 (9.8 per cent) over 50 per cent. In this analysis internationalisation of creative activity seemed most pronounced for UK firms, which accounted for only 3.7 per cent of the enterprises with no overseas patents and 3.0 per cent of those with an overseas subsidiary patent ratio of less than 10 per cent, but for 36.8 per cent those with one of over 30 per cent. By contrast US firms accounted for 41.9 per cent of those without patents of overseas origin, and for 66.2 per cent of enterprises with a ratio of less than 10 per cent, but for only 7.7 per cent of those with a ratio of over 30 per cent. Of 124 Japanese companies in the analysis only 3 (2.4 per cent) had an overseas subsidiary patent ratio in excess of 10 per cent, whilst it was zero for 91 (73.4 per cent).

For 479 of the companies for which the overseas subsidiary patent ratio was available we also had information on the extent of overseas production, in the form of the 'overseas production ratio' (i.e. 'overseas production as a percentage of total group sales')[6]. Once the 87 firms with an overseas production ratio of less than 5 per cent were excluded it was found that only 68 (17.3 per cent) of the remainder had an overseas subsidiary patent ratio of zero, with 163 (41.6 per cent) having a positive ratio of up to 10 per cent, 82 (20.9 per cent) between 10 per cent and 30 per cent, 33 (8.4 per cent) between 30 per cent and 50 per cent and 46 (11.7 per cent) over 50 per cent. Finally for the 193 MNEs with an overseas production ratio of over 25 per cent only 12 (6.2 per cent) had a zero overseas subsidiary patent ratio whilst 57 (29.5 per cent) had a positive ratio of less than 10 per cent, 58 (30.1 per cent) between 10 per cent and 30 per cent, 25 (12.9 per cent) between 30 per cent and 50 per cent and 41 (21.2 per cent) over 50 per cent. Generally these results tend to illustrate the emergence of globalised creative activity in MNEs in support of globalised markets and production.[7]

The context for the increasing internationalisation of R&D by MNEs that we have outlined implies that this may be stimulated by supply and/or

---

[5] These derive from the sample of 792 leading industrial enterprises in the world in 1982 analysed by Dunning and Pearce [3]. Details of the assembly of this 792 firm sample are given in [3, pp 8-11].
[6] For sources of the data used to derive the overseas production ratio, see [7, p. 90].
[7] For formal tests of the determinants of overseas R & D using this data, see [9].

demand side influences. On the one hand a MNE's growing knowledge of international environments, through overseas marketing and production, may extend to a recognition of distinctive scientific capabilities which could serve specialised needs in its research programme. Alternatively the same experience of local environments may more directly point to a certain degree of incompatibility between the MNE's existing products and processes and local needs. R&D work may then be carried out locally to adapt or develop MNE products and processes in ways that enhance their suitability to indigenous consumer requirements and productive potentials. It is an underlying contention of our work that these two (supply and demand side) influences on internationalisation of MNE R&D need not match up in any given national context. The most distinctive scientific capabilities accessible to MNEs in a country may exceed those needed for adaptation or development work, and may be more valuably harnessed into the more ambitious creative programmes of the group. By contrast the locally-focused adaptation or development work can often be performed by routinely capable members of the local scientific labour force in conjunction with expatriate MNE personnel. In addition to these supply and demand side factors our background perceptions also imply that overseas R&D may be responding more directly to particular competitive needs of the MNE. A particular facet of this may relate to possible aspects of oligopolistic competition within R&D (i.e. a perceived need to match certain R&D moves of rival firms).

Against this conceptual background our research methodology distinguishes three types of R&D performed in overseas laboratories. The first of these involves R&D undertaken to adapt an existing product, or production process, to the needs of a particular host country. We refer to R&D units that are predominantly committed to this type of work as Support Laboratories (SLs), since it is their function to support the efficient local assimilation of the MNE's current technology. In terms of the four approaches to innovation in MNEs distinguished by Bartlett and Ghoshal, the work carried out by SLs may be seen as playing a role in support of their 'centre-for-global' strategy, by facilitating the effective world-wide exploitation of a new product or process originally created through the use of the centralised resources of the parent company. Bartlett and Ghoshal [1] indicate that where used a 'centre-for-global' approach to innovation is "necessary because certain key capabilities of the MNE must, of necessity, remain at the headquarters both because of the administrative need to protect certain core competencies of the company, and also to achieve economies of specialisation and scale in the R&D activity". This perspective suggests three of the established arguments against substantial diffusion of R&D capability in MNEs; loss of efficiency due to communications problems, a concomitant danger of increased risk of losing control over key knowledge before its adequate development and exploitation, failure to realise economies of

Internationalisation of R&D: Organisation and Motivation ———— 141

scale in valuable scientific inputs (personnel or machinery). It should also be observed that the procedures of the 'centre-for-global' innovation process closely parallel the geographical market-spread, and subsequent overseas production implications, of the traditional product cycle [11]. The implications of this product cycle for R&D location in MNEs have also been discerned[8]. Thus the basic research, and innovation-supporting development, will be totally the responsibility of the parent company, with certain amounts of product and process adaptation work carried out by Support Laboratories to effect the later geographical diffusion of sales and production.

The second type of overseas R&D distinguished in our work is that which seeks to *develop* new products or processes which go beyond mere assimilation of products derived elsewhere in the MNE group. Because effective performance of this type of work is predicated upon feedback from local marketing personnel and engineers, the unit involved is characterised as a Locally Integrated Laboratory (LIL). An important variant of the work of a Locally Integrated Laboratory occurs when it functions within a subsidiary that has been given a world (or regional) product mandate (WPM)[9]. In this case the subsidiary has been endowed with the responsibility for deriving, producing and marketing a distinctive product in the MNE's range for the global (or at least a large regional) market. The LIL, though probably still operating within the broad confines of the MNE group's existing technology, works with subsidiary marketing and production/engineering personnel to derive the product to fulfil this mandate.

Locally Integrated Laboratories may be seen to play roles in two of the innovation strategies distinguished by Bartlett and Ghoshal, these being the second of the two 'traditional' strategies (along with 'centre-for-global'), i.e. 'local-for-local', and the first of the two emergent 'transnational' strategies, i.e. 'locally-leveraged' innovation. In the 'local-for-local' approach national subsidiaries of MNEs use 'their own resources and capabilities to create innovations that respond to the needs of their own environment' [1], The traditional genesis of 'local-for-local' innovation reflected the effective isolation of important markets and the need to react to the persistence of notably distinctive characteristics in them. More recently the emphasis, whilst still acknowledging the need to respond to distinctive local market needs, has moved towards the competitive need for a very quick international diffusion and differentiation of innovations. With the perception[10] that in some industries the product cycle has become compressed 'into something resembling a programme of near simultaneous innovations in several markets' [7] the role for LILs in supporting innovation strategies

---

[8] See [7, pp. 5-6].
[9] See [8, pp. 121-130].
[10] Acknowledged by Vernon [10].

may have become one of liaising between the evolving central technology underlying innovation and the need to impart strong differentiated local characteristics to its dispersed implementation. As already observed LILs also seem likely to provide the scientific inputs into 'locally-leveraged' innovation, which as defined by Bartlett and Ghoshal [1] involves "utilizing the resources of a national subsidiary to create innovations not only for the local market but also for exploitation on a world-wide basis", i.e. in support of World Product Mandate subsidiaries.

The third type of overseas R&D in our typology is that which plays a role in the *long-term* basic research programme of the group. When a MNE feels it has located a source of scientific capability whose competence exceeds the demands of SL or LIL work it may incorporate it into a wider programme of more basic research focused on the derivation of future generations of products. Units performing this type of work are described as Internationally Interdependent Laboratories (IILs). An IIL participates in an R&D initiative which is likely to involve several similar units in other locations. The IIL is unlikely to be significantly involved with producing and marketing units in the same host country, and, therefore, there is no reason to expect the results of its work to directly benefit them. Our Internationally Interdependent Laboratories parallel the 'globally-linked' approach to innovation discussed by Bartlett and Ghoshal [1] which 'pools the resources and capabilities of many different components of the MNE—at both the headquarters and the subsidiary level—to create and implement an innovation jointly. In this process, each unit contributes its own unique resources to develop a truly collaborative response to a globally perceived opportunity'.

7.2 THE SURVEY

The sample of parent or overseas subsidiary laboratories to which our questionnaires were sent derives from 560 major enterprises. Within each enterprise it was our aim, where relevant, to survey both the parent and all its subsidiary laboratories.

Parent laboratories are either corporate level R&D units (either physically located at corporate headquarters, or otherwise distinguished as the corporate unit) or the main R&D facilities of major divisions in diversified enterprises. It was our *a priori* expectation, subjected to investigation in the questionnaires, that such units may (a) perform the more basic R&D the results from which may provide the basis for development by subsidiary laboratories, including those overseas, and (b) conceive, initiate and coordinate integrated research programmes which may involve, amongst others, overseas laboratories.

Overseas Subsidiary Laboratories are non-parent units located outside the home country of the group. The laboratories may support local engineering or marketing needs, or make a specialised contribution to a globally inte-

Internationalisation of R&D: Organisation and Motivation — 143

grated research programme originally articulated and implemented by a parent unit.

The starting point in the derivation of the 560 enterprises underlying our samples was the 500 largest industrial enterprises in the world in 1986 as derived from the Fortune listings. A number of the leading directories of R&D facilities were scrutinised to find the parent and overseas subsidiary laboratories of these 500 companies [2,4,5,6]. Since large divisionalised companies were often considered to have more than one parent R&D unit, more than 500 parent laboratories were found for these firms. On the other hand, not all had overseas facilities and the total of these came to less than 500[11]. In reviewing the Directories to locate facilities for the 500 largest enterprises a number of companies outside this sample were discovered which seemed to constitute relevant cases of decentralised R&D. A sample of 30 of these companies was distinguished and their parent and subsidiary units surveyed. Since these 30 extra firms constituted a self-selecting sample of firms with some commitment to international R&D, a further group of 30 enterprises were selected which had no apparent overseas R&D operations. These were chosen to match the other 30 by industry and home country as closely as possible.

A total of 1028 questionnaires were sent out, and 296 usable replies were received, an overall response rate of 28.9%. Parents supplied 163 of the responses, and subsidiaries 133. The response rate was slightly higher for subsidiaries than for parents, i.e. 32.8% compared with 26.2%.

7.3 RESULTS

The survey results support the emerging perception of a growing role for internationalised R&D in MNEs. Whilst overseas R&D remains in the aggregate a relatively small proportion of the total carried out by industrial enterprises, it is also shown to have taken a significant position in the creative and competitive operations of many major MNEs. Thus the results indicate the manner in which overseas R&D now plays a number of clearly defined and carefully articulated roles in support of the global competitive strategy of leading enterprises.

With respect to the extent of global dispersion of R&D activity by the major enterprises in our survey, it was found[12] that of 114 parent laboratory respondents to a question on the percentage of their group R&D expenditure carried out by their overseas R&D units 44% reported no such expenditure and another 13% less than five%. Though it is likely that many of these companies also had very limited overseas production and sales this still makes

---

[11] To cover cases with no overseas R&D a section of the questionnaire sent to parent laboratories was directed to such enterprises, seeking elucidation of the lack of such decentralisation.
[12] See [9, Table 3.1].

it clear that globalisation of R&D is not yet a widely pervasive practice. Nevertheless 19% of respondents reported the commitment of between 20% and 50% of their group budgets to overseas laboratories' R&D, and 5% over 50% of expenditure. Internationalised R&D was found to be well established in UK enterprises (52% reporting over 20% of expenditure overseas) and those from Other European companies (47%), whilst the below average tendency of US firms to use such operations is reflected in the comparable figure of 14% and also in the 74% (compared with 66% for all respondents) with less than 10% of expenditure overseas. Though there are clear indications of a fast-growing interest in overseas R&D amongst globally competing Japanese enterprises this has not yet reached strong quantitative levels, with no respondents reporting over 5% of group budgets overseas and 71% no such expenditure. At the industry level the global perspective on R&D is most clearly established in pharmaceuticals, where 63% of respondents reported over 20% of expenditure overseas, and least in aerospace (where overseas production is also notably restrained) and metal manufacture and products.

Whatever the current situation with respect to aggregate levels of overseas R&D, our evidence supports the view that reformulated attitudes to innovation in enterprises competing through a global business strategy require, and are receiving, the support of dispersed R&D operations playing an expanded range of roles in technology creation and its efficient commercial application. Thus information supplied by the parent laboratories on the ages of their group laboratories[13] shows that only 27% of laboratories set up before 1970 were located abroad, whilst 65% of those set up since 1970 were overseas. In addition evidence derived from the subsidiary questionnaire indicated the relative frequency amongst recently established overseas laboratories of those that considered themselves as predominantly Internationally Interdependent Laboratories and that regularly performed basic/original research[14]. This reflects the view that overseas facilities are increasingly set up to play specialised roles in globally-oriented R&D programmes, and that this is increasingly likely to involve more ambitious types of quite fundamental research, which may in turn reflect the acknowledgement of the value of distinctive scientific capabilities in dispersed foreign locations. Each of the themes suggested by the results noted above was more directly investigated elsewhere in the two questionnaires.

Thus further support for certain of the perspectives emerged from a question in which parent laboratories were asked how they evaluated possible changes in the international location of R&D in their company, in terms of

---

[13] See [9, Table 3.2].

[14] The subsidiary questionnaire distinguished six types of research. Basic/original research; applied research to derive new products in the present industry; applied research to derive new production technology in the present industry; applied research to adapt existing products to the local market; applied research to adapt existing production technology to the local environment; applied research to derive additional products in new areas of specialisation.

Internationalisation of R&D: Organisation and Motivation — 145

four offered alternatives. The dominant response, selected in 66% of replies[15], was that they expected to place 'increased emphasis on a globally integrated R&D network'. Japanese companies (81% of their respondents) showed a notable enthusiasm for this approach, which, in view of the particularly recent origins of most overseas R&D in these enterprises, indicates a very prompt implementation of this sophisticated approach to its organisation. Our investigations also showed that parent laboratories regularly doing basic/original research were distinctly most prone to adopt this approach to the possibilities of global R&D work. This in turn is compatible with a motivation to disperse R&D work in order to make the most effective use of sources of specialised capability located in foreign countries. Further, another 9% of replies said that they planned 'more emphasis on autonomous overseas laboratories', thus augmenting the apparent momentum towards increased use of foreign-based facilities, though not their systematic networking. Against this 20% of replies indicated a likely preference for 'more use of centralised facilities'. That only 5% of respondents felt that 'no change' in policy was imminent endorses the overall view of the globalisation of R&D being a prevalent contemporary issue in leading MNEs.

Elsewhere in the two questionnaires other results provide further support for the view that the majority of overseas R&D facilities now play roles that subject their operations to consistent review, if not active coordination, by parent laboratories. Thus 68% of parent laboratories with overseas subsidiaries selected 'systematic coordination' as their prevalent form of interaction with them, while another 25% considered '*ad hoc* consultation' to describe their involvement with the subsidiaries' work, and only 7% felt interaction was 'infrequent'. However, when subsidiary laboratories were asked their perception of the parent's strategy towards such units 38% felt that allowing 'substantial autonomy' was the prevalent approach, and 28% that it was 'incorporating subsidiary work into carefully coordinated programmes'. Between these extremes 33% of subsidiary laboratory respondents felt that 'allowing subsidiary units to develop independent initiatives, but under close central supervision' best described the parent's approach[16].

---

[15] The question was addressed to all parent laboratories including those without overseas R&D at the time of reply. Of the 166 parent laboratories responding only 119 answered this question. Obviously it is likely that the majority of non-respondents are those without overseas R&D, and who do not currently anticipate implementing it. Nevertheless the results do also imply the planned implementation of overseas R&D in respondents currently without it.

[16] In fact several other results obtained elsewhere in the subsidiary questionnaire seem most compatible with an approach in which many parent laboratories derive their global operations by balancing and coordinating a programme of ideas proposed by subsidiary units, rather than by an exclusive reliance on the imposition of their own, centrally generated, ideas. More generally an earlier analysis of world product mandate subsidiaries [8, pp. 15–16] indicated the same approach to coopting distinctive creativity. Thus 'the more an MNE moves towards a willingness to accommodate WPMs the more it is moving to a situation where central planning balances current initiatives and constructs the overall growth programme from such initiatives, rather than by using a central view of a unique path and building subsidiary potentials into this vision'.

Table 7.1 Nature and frequency[a] of parent or sister affiliate involvement in the projects of subsidiary R&D units

	Type in involvement (average frequency[b])			
	Systematic coordination of projects into wider programmes	Intervention to bring about a major change in the direction of the project	Advice on the development of the project	Technical assistance at the request of the R&D unit
Industry				
Food, drink, tobacco	1.91	1.64	2.18	2.55
Petroleum	2.50	2.40	2.40	2.60
Metal manufactures and products	1.91	1.55	1.73	2.09
Industrial and agricultural chemicals	2.28	1.78	2.11	2.08
Pharmaceuticals and consumer chemicals	2.60	2.00	2.47	2.33
Motor vehicles and components	2.50	1.25	2.25	2.00
Industrial and farm equipment	2.50	1.75	2.25	2.25
Office equipment (incl. computers)	2.22	1.78	1.89	2.11
Other manufacturing	2.33	2.00	2.17	2.00
Total	2.31	1.82	2.17	2.23
Host country				
USA	2.19	1.70	2.18	2.23
UK	2.43	2.09	2.18	2.21
Other Europe	2.33	1.67	2.11	2.33
Japan	3.00	3.00	2.50	2.00
Other Countries	2.44	1.78	1.89	2.11
Total	2.31	1.82	2.17	2.23
Home country				
USA	2.36	1.83	2.17	2.23
UK	2.05	1.70	2.15	2.15
Other Europe	2.45	1.88	2.27	2.29
Japan	1.25	1.50	1.50	2.00
Other Countries	2.33	2.00	1.67	2.33
Total	2.31	1.82	2.17	2.23

[a] Respondents were asked to grade frequency of the variuos types of involvement on the scale 1: Never, 2: occasionally, 3: regularly.
[b] The average derived by allocating values to the responses of 1 for 'never', 2 for 'occasionally', 3 for 'regularly'.

In a complementary question subsidiary laboratory respondents were asked to assess the extent of various forms of involvement in their projects by the parent or other R&D units in their MNE group (see Table 7.1). With regard to 'systematic coordination of projects into wider programmes', 41% of subsidiaries felt this was a 'regular' role for their projects, 49% that it happened 'occasionally' and only 10% felt their projects 'never' played this role. The greater degree of central influence implied here may reflect a difference between subsidiary respondents' evaluation of their own (greater) subjection to central scrutiny/coordination and that which they perceive as the norm in the group. Also the higher response to coordination in this question may reflect joint programmes with other subsidiary units, which are not subjected to central intervention. In a more general question 75% of responding subsidiary laboratories felt they interacted with other group units 'very often' and only 2% 'never'.

A practice that would exemplify the use of global coordination in pursuit of the optimal completion of a piece of R&D work through the best use of dispersed specialised abilities is the mobility of projects, so that research started in one laboratory might be transferred to another for completion. Our investigations indicate that this does play a role in global R&D as implemented by parent laboratories. Thus 26% of respondents felt that transferring projects *from* overseas units to the parent was a frequent or automatic procedure, and only 18% indicated it never happened. Also 44% of these parent respondents said that transfer of their projects to overseas subsidiary laboratories was a frequent or automatic occurrence, and only 9% that it never happened. The prevalent motives for these transfers reflect the expected specialisations and roles of these units. Thus the strongest reason for moving projects to the parent from overseas units (55% of replies) was that the central facility could 'better complete the research work'. By contrast the dominant motive for the movement of projects from the parent to subsidiaries (81% of replies) was to 'ensure an outcome best directed to a particular market'.

With clear indication of an expanded range of possible roles for overseas R&D in MNEs, and with the suggestion that many such units have therefore developed specialised functions, often within globally-integrated group programmes, we turn to review the evidence on the relative prevalence of the various laboratory types and of the different types of R&D that are performed in them[17]. The increased contribution that overseas laboratories expect to make to the sustained competitiveness of their MNE groups is reflected in their limited evaluation of themselves as Support Laboratories, i.e. playing the traditional role of *adapting* existing products and/or processes to host country conditions. Thus only 22% of respondents that evaluated themselves by SL criteria believed that they 'predominantly' played this role and 37% that they 'partially' did. Clearly, however, one strand in the increased ambition of overseas R&D has been an evolution of this local support role, in the form of Locally Integrated Laboratories which seek to *develop* new products and/or processes to meet distinctive needs of particular environments in which the group seeks to sustain competitive operations. Of the respondents that evaluated themselves by LIL criteria

---

[17] See [9, Tables 5.1 and 5.2] for more detail.

30% said they were 'predominantly' of this type and 45% 'partially'. Finally the perception of overseas R&D being increasingly required to play its specialised roles in the direct service of broader group needs, through involvement in integrated programmes, is supported by the degree to which these units evaluated themselves as Internationally Interdependent Laboratories. Thus of the respondents that indicated their IIL status, 44% said they were 'predominantly' that type, and 30% 'partially'.

Turning to the way in which the overseas laboratories assessed their commitment to various types of research it becomes clear that, despite their frequent elevation to well defined specialised roles within MNE operations, their predominant function remains the support of the commercial implementation and evolution of the group's technology. Thus 60% of responding subsidiary laboratories said they 'regularly' (as distinct from 'occasionally' or 'never') did work to adapt existing products to local markets and 50% regularly sought to derive new products in the present industry. With respect to work aimed at production processes there was a rather surprising difference of emphasis. Here 74% regularly did applied research to derive new production technology in the present industry, compared with 42% regularly doing work to adapt existing production technology to the local environment. Finally 44% of the overseas subsidiary laboratories reported they regularly did research aimed at deriving additional products in new areas of specialisation.

It is an implication of these results that many overseas subsidiary laboratories have a regular commitment to the performance of more than one type of R&D, possibly involving work on both products and processes at the adaptation and/or development level. What is also implied, by the balance (already noted) between units describing themselves as predominantly Support Laboratories or Locally Integrated Laboratories is that where laboratories do encompass a range of these types of research they usually feel themselves to be most clearly defined by their development role, with adaptation a regular but less central commitment. The persisting dominance of adaptation and/or development work in these overseas R&D facilities, taken in conjunction with the prevalence of Internationally Interdependent Laboratories amongst the types of unit, indicates that an important aspect of the evolution of these laboratories is often to provide their services to support production or marketing operations outside their host country.

Of course the dominant position taken by adaptation and development in the operations of these overseas laboratories to quite a substantial degree merely reflects the preponderant role of these stages in creative operations in general, with basic/original research also being a minority activity in the parent laboratories. Nevertheless our results make it clear that in relative terms basic/original work is least prevalent in the overseas units and that their distinctive focus is still on the development and adaptation stages of

Internationalisation of R&D: Organisation and Motivation ——— 149

the commercial exploitation of technology. Thus of 79 responding parent R&D units 30% did not carry out basic/original research in either the parent unit or any overseas unit. However, where such work was done it tended to be most strongly focused on the parent units, with 57% believing this type of work to be relatively less important in overseas units than the parent, 20% equally important in both types of unit, whilst only two respondents believed it to be relatively more important in subsidiary units. By contrast, for each of three types of adaptation or development research investigated, over half the responding parent laboratories believed these types of work to be equally important in parent and subsidiary units[18].

Reflecting the balance in the operations of overseas units already noted only 17% of responding subsidiary laboratories felt they did basic/original research regularly, though a relatively high 47% did do such work occasionally. Regular performance of basic/original work in these laboratories is also shown[19] to be a relatively isolated activity, rarely juxtaposed with regular performance of other types of R&D. It is also interesting to note that of the respondents that described themselves as IILs and also defined the nature of their work only 29% did basic/original research regularly. This tends to play down, as earlier results have also already suggested, our original perception of internationally integrated operations in MNEs' global R&D activity as being primarily focused around building distinctive host country capabilities into basic research programmes seeking the original scientific perspectives from which applied R&D can derive innovative commercial breakthroughs. However, it should be noted that of the overseas laboratories that regularly did basic/original work and specified their laboratory types 84% were predominantly IILs. By indicating that where basic/original work is done in subsidiary laboratories it is overwhelmingly oriented towards a role in wider programmes this reinforces the complementary perspective that the development work which is increasingly prevalent in these units rarely represents the completion of their own basic/original inputs. From the point of view of countries acting as hosts to MNE R&D facilities these results may be seen as having the doubly worrying implication that local resources applied to basic/original work produce results that leave the country, with no potential for generating direct benefits in local productive output, whilst the development and adaptation work remains dependent on the assimilation of externally generated technology.

In the light of the results reported on laboratory types and the nature of the work done in them, it is not surprising that investigation of the factors influencing recent decisions affecting their development (see Table 7.2) indicates a predominant role for 'demand-side' factors (i.e. those relating to the

---

[18] See [9, Chapter 3.4] for more detail.
[19] For more detail, see [9, Table 5.3].

Table 7.2(a) Conditions and circumstances considered to have most influenced[a] recent decisions with regard to development of subsidiary R&D units by industry

Types of influence: Average response[b]	Food drink tobacco	Petroleum	Metal manufacture & products	Industrial & agricultural chemicals	Pharmaceuticals & consumer chemicals	Motor vehicles (including components)	Industrial & farm equipment	Electronics & electrical appliances	Office equipment (including computers)	Other manufacturing	Total
A distinctive local scientific, educational or technological tradition conducive to certain types of research	1.60	2.00	1.20	1.83	2.07	1.00	1.50	1.83	1.86	1.67	1.78
Presence of a helpful local scientific environment and adequate technical infrastructure	1.54	2.17	1.18	1.65	2.03	1.40	1.50	1.77	1.75	1.83	1.73
Availability of research professionals	1.82	2.50	1.46	2.05	2.20	1.80	1.50	1.92	1.88	2.17	1.99
Favourable wage rates for research professionals	1.27	1.50	1.46	1.42	1.53	1.80	1.75	1.31	1.25	1.33	1.44
Need to provide technical services to local production unit	2.46	2.17	2.36	2.22	1.77	3.00	2.50	2.21	1.50	2.33	2.14
To help modify/standardise products for the local market	2.82	2.00	2.36	2.69	2.07	2.50	2.25	2.38	1.37	2.17	2.34
To help modify/standardise products for overseas markets	2.36	2.17	2.00	2.11	1.79	1.75	2.00	2.00	1.75	2.00	1.97
To help develop new products for the local market	2.91	2.33	2.82	2.73	2.47	3.00	2.50	2.31	2.38	2.50	2.60
To help develop new products for overseas markets	1.91	2.33	2.36	2.19	2.34	2.25	2.25	2.15	2.75	1.83	2.24
To provide technical support to other parts of the multinational group	1.91	2.83	1.91	2.14	2.03	2.75	1.76	1.77	1.87	1.83	2.05
A large and growing local market where R&D is seen to play a critical role	1.91	2.00	1.46	1.97	2.30	2.26	1.50	1.92	1.50	2.00	1.96
To forestall entry of another firm	1.18	1.00	1.00	1.19	1.07	1.00	1.00	1.17	1.00	1.17	1.11
To match local R&D of competitor firm	1.46	1.33	1.27	1.44	1.31	2.00	1.50	1.67	1.25	1.17	1.41

[a] Respondents were asked to grade each condition or circumstance on the scale, (1) irrelevant to decisions, (2) of some influence on decisions, (3) a major factor contributing to decisions.
[b] The average derived by allocating values to the responses of (1) for 'irrelevant', (2) for 'some influence', (3) for 'major factor'.

Table 7.2(b) Conditions and circumstances considered to have most influenced[a] recent decisions with regard to development of subsidiary R&D units by host country and home country

Types of influence: Average response[b]	Host Country					Home Country			
	USA	UK	Other Europe	Japan	Other countries	USA	UK	Other Europe	Japan	Other countries
A distinctive local scientific, educational or technological tradition conducive to certain types of research	1.76	1.83	1.71	2.50	1.78	1.71	1.76	1.92	1.50	1.33
Presence of a helpful local scientific environment and adequate technical infrastructure	1.67	1.71	1.68	2.50	2.11	1.64	1.62	1.90	1.50	1.33
Availability of research professionals	1.90	2.12	1.96	2.50	2.11	2.02	1.91	2.04	1.50	2.00
Favourable wage rates for research professionals	1.38	1.64	1.29	1.00	1.67	1.45	1.52	1.39	1.25	1.67
Need to provide technical services to local production unit	2.27	1.88	2.11	2.00	2.33	2.08	2.48	2.10	1.75	2.00
To help modify/standardise products for the local market	2.51	2.03	2.44	2.00	2.22	2.25	2.57	2.31	2.75	2.33
To help modify/standardise products for overseas markets	1.79	1.94	2.20	1.50	2.33	2.14	1.91	1.80	2.00	2.33
To help develop new products for the local market	2.76	2.53	2.48	2.00	2.33	2.56	2.86	2.58	2.00	2.67
To help develop new products for overseas markets	2.03	2.68	2.20	1.50	2.22	2.44	2.05	2.14	1.50	3.00
To provide technical support to other parts of the multinational group	1.85	2.29	2.15	2.00	2.11	2.21	1.86	1.98	1.00	3.00
A large and growing local market where R&D is seen to play a critical role	1.96	1.88	1.93	2.50	2.22	1.92	2.24	1.98	1.25	1.33
To forestall entry of another firm	1.09	1.09	1.15	1.50	1.11	1.11	1.35	1.02	1.00	1.00
To match local R&D of competitor firm	1.57	1.21	1.27	2.50	1.33	1.21	1.65	.157	1.00	1.00

[a] Respondents were asked to grade each condition or circumstance on the scale, (1) irrelevant to decisions, (2) of some influence on decisions, (3) a major factor contributing to decisions.
[b] The average derived by allocating values to the responses of (1) for 'irrelevant', (2) for 'some influence', (3) for 'major factor'.

service the laboratory is to provide to other parts of the group, rather than those affecting its ability to provide the service or as a response to aspects of the firm's technological, or wider, competitive situation). Thus 45 % of respondents considered that the 'need to provide technical services to the local production unit' had been a 'major factor' contributing to recent decisions on the laboratory's development (as distinct from being 'of some influence' or 'irrelevant'). Also 32% rated the provision of 'technical support to other parts of the MNE group' to be a major influence, backed up by a notably high 41% who felt it to be of some influence[20]. Work to 'help modify or standardise products for the local market' was rated as a major factor influencing decisions by 54% of respondents, whilst this work for other markets was similarly rated by 33% (though a particularly high 35% found it irrelevant). Finally work to 'develop new products' for the local market (70% of respondents rating it a major factor) and for other markets (45%) was the most pervasive influence of all on overseas R&D unit decision making.

A further important result distinguished in the analysis of these demand-side influences concerns the position of IILs. It emerges that whereas IILs are consistently the least likely type of laboratory to consider local demand-side factors a major influence on their development they are relatively very much more prone to provide this rating for overseas demand factors. Thus for technical services to the local production unit 73% of SLs, 53% of LILs and 30% of IILs rated this a major influence on their decisions. The comparable figures for technical services to other parts of the group were 31%, 25% and 37%. In the case of work to 'modify or standardise the product' 85% of SLs rated this a major factor for local markets but only 40% for overseas markets, whilst the comparable figures for LILs were 75% and 28% and for IILs 32% and 34%. Similarly for work to develop new products 81% of SLs rated this a major factor for local markets but only 28% for overseas markets, with comparable figures of 89% and 36% for LILs and 54% and 57% for IILs. This further supports the view that IILs have broader roles in MNE networks than the originally perceived one of coopting distinctive local scientific capability into basic/original research oriented global programmes.

These results therefore indicate that MNEs have internationally-integrated perspectives on the supply of adaptation and development work, as well as of basic/original research. Whilst their experience in international cooperation and communications is obviously central to IILs' relative prevalence in performance of this role, their hypothesised access to higher levels of more distinctive technical expertise may also be a contributory factor. In fact the main basis of their international interdependence may still be within

---

[20] In fact while only 27% of respondents felt 'external' technical support to be irrelevant to decisions, 31% felt local technical support was irrelevant.

Internationalisation of R&D: Organisation and Motivation — 153

the research network if the adaptation and development problems they are asked to address are defined for them by SLs or LILs in the countries requiring support, and passed on when these units acknowledge the need for specialised assistance. There is some support for this from analysis of laboratories performing different types of research. Thus laboratories regularly performing basic/original research rated the meeting of external needs more often a major influence than local needs for all three services (technical support; product modification; new products), whilst the opposite was the case for laboratories regularly doing other types of research. Nevertheless basic/original remained least influenced, even from external needs, for all except 'new products', where it did become most likely to see provision of external services as a major influence.

Turning to supply-side factors, the question (see Table 7.2) investigating the factors influencing the recent decisions affecting the development of overseas laboratories in MNEs incorporated several of these. The response to 'a distinctive local scientific, educational or technological tradition conducive to certain types of research project' was less than expected. Thus 23% of respondents rated this as a major factor influencing their decisions, and 44% considered it irrelevant. More in line with expectations 29% of IILs (compared with 9% of LILs and 8% of SLs) considered it a major influence, as did 38% of laboratories regularly doing basic/original research (compared with between 10% and 23% for the other five types of research). In fact respondents seemed reluctant to distinguish between the relevance of these distinctive or idiosyncratic capabilities of the host-country technological environments and the more general level of abilities offered. Thus replies to the 'presence of a helpful local scientific environment and adequate technical infrastructure' were very similar to those for the previous condition, both in the aggregate and with respect to the stronger response from IILs and laboratories regularly doing basic/original research. Of two specific conditions within the technological environment that were investigated, 'availability of research professionals' emerged as slightly more influential than the assessments of the more widely defined characteristics. Thus 28% of respondents rated it a major factor influencing their decisions, including 37% of those that were predominantly IILs and 55% of those regularly doing basic/original work. By contrast, no respondents reported that 'favourable wage rates for research professionals' had been a major factor in their decisions, whilst 62% deemed it irrelevant[21].

When parent laboratories evaluated the influence of a number of factors on the type of work done in their overseas R&D units they also found lim-

---

[21] Industries in which supply-side factors are of relatively high importance are office equipment (including computers), pharmaceuticals and consumer chemicals, and petroleum. Also foreign laboratories in the UK rated supply-side factors highly, as did the overseas laboratories of Other European MNEs (see Table 7.2).

ited relevance for supply-side factors. Thus 15% of respondents rated 'a distinctive local scientific, educational or technological tradition conducive to certain types of project' as 'nearly always relevant' and 32% 'never relevant'. 'Cost factors' in general were also rated as nearly always relevant by 15% of respondents and never relevant by 29%. The types of work done by these parent laboratories had no influence on how they valued these influences on their overseas subsidiaries. Also, amongst a subset of parent laboratories that had considered but not implemented overseas R&D, 24% felt that the 'desire to incorporate foreign located sources of expertise in centrally coordinated international research programmes' had been the key factor stimulating this interest. In the same question 25% of respondents had felt that the 'desire of foreign producing facilities to upgrade their technological capability' had been the key influence, though this may be mainly perceived as a demand-side factor. With respect to the reasons for the ultimate rejection of overseas R&D by parents that had considered it, evaluation of relative supply-side capability seemed of quite limited relevance, with 32% indicating that 'no overseas locations have the expertise to rival the home country units' had been a major cause of rejection but 58% considered it irrelevant. However, amongst parent laboratories that had not actively considered overseas R&D, 60% of respondents felt a major factor ruling out such consideration was that 'the home country research environment, including skills of scientists, is fully adequate for our needs'.

Several characteristics of the process of technology creation have been alleged to mitigate against its safe and efficient diffusion into a number of dispersed R&D sites. These may be evaluated here as an extension of the supply-side factors. The most frequent line of argument of this type indicates that extensive economies of scale exist in R&D and that the desire to fully realise these, most often in a parent laboratory, constrains enthusiasm for geographically-dispersed operations. Since the assessment of such a view requires the parent perspective, both to evaluate the extent of such economies of scale per se and its relevance to decisions on the geographical balance of group creative activity, it was investigated in the parent questionnaire. When parent laboratories that did have overseas R&D were asked to evaluate the influence of various factors on types of work done in their overseas facilities the possibility that there was 'only room for a small number of basic R&D laboratories' was considered to be nearly always relevant by 29% of respondents and never relevant by 48%. When parents that had considered but rejected overseas R&D were asked to evaluate the influence of 'research economies of scale requiring centralised facilities' 33% said this had been a major cause of rejection and only 24% that it had been irrelevant. Finally 48% of parent laboratories that had not actively considered overseas R&D rated 'scale factors must limit our research to one site' to be a major factor ruling out consideration of foreign R&D and only 21% that it

Internationalisation of R&D: Organisation and Motivation 155

had been of no influence. It would thus seem, to some degree, that the closer decision makers are to actual experience of overseas R&D the more economies of scale influences recede in their practical evaluation[22].

The sensitivity of research outputs has often prompted the view that R&D would be strongly centralised, in order to maximise their security. There is only limited support for this belief in our results. Amongst parent laboratories that had considered but rejected overseas facilities 16% said that 'the sensitivity of our research results requires close home country control' was a major cause of rejection, and 63% that it was irrelevant. Similarly of parent respondents that had not considered overseas R&D only 13% rated 'sensitivity of our research' as a major factor ruling out its consideration and 58% felt it had been of no influence. Finally the view that 'communications problems with dispersed units would harm the types of R&D we do' was substantially repudiated by the parent laboratories that had considered but rejected such overseas facilities. Thus 21% felt this a major cause of rejection and 53% rated it as irrelevant.

Another essentially supply-side area of investigation relating to these overseas subsidiary laboratories is the extent to which their work may be considered to be science-driven, in the sense of implementing ideas derived within the science communities of the group, either in response to its own basic/original research or as a result of a distinctive perception of publicly accessible knowledge in the relevant areas of science. Some evidence on this emerges from the replies of subsidiary laboratories to a question which asked them to evaluate a number of possible sources of the project ideas that they initiate (see Table 7.3). Only 13% of respondents believed that suggestions from their parent laboratories were a regular source of ideas, though a further 70% acknowledged that they occasionally were. Though this source of ideas was, as expected, most often a regular input into IILs (24 per cent of IIL respondents compared with 12% of SLs and 3% of LILs) its influence there was much less than anticipated. One reason for this may be that parent laboratories may be seeking to tap into the creative ideas of IILs as well as their scientific skills, and thus seek to balance a global programme of initiatives rather than impose only their own conceptions. Also the low rating of parent idea inputs in IILs may reflect the higher than expected propensity for these units to provide adaptation and development work in support of operations in third countries. Only 3% of respondents, however, felt that their project ideas were regularly suggested by sister R&D affiliates, though 61% felt this was an occasional source. In the light of the quite extensive commitment of these laboratories to do adaptation or

---

[22] Pearce [7, pp. 38–40] provides reasons why economies of scale factors may become less influential on the evaluation of overseas R & D when this is already clearly entrenched in an MNE's operations, compared with where this is only under consideration.

Table 7.3(a) Sources of project ideas initiated in overseas subsidiary R&D units[a], by industry

Sources of ideas: Average response[b]	Food drink tobacco	Petroleum	Metal manufacture & products	Industrial & agricultural chemicals	Pharmaceuticals & consumer chemicals	Motor vehicles (including components)	Industrial & farm equipment	Electronics & electrical appliances	Office equipment (including computers)	Other manufacturing	Total
Suggested by parent	1.73	2.20	1.73	2.00	2.13	2.25	2.00	1.92	1.75	1.67	1.97
Suggested by sister affiliates	1.64	2.00	1.73	1.81	1.57	1.80	1.50	1.58	1.38	1.67	1.67
Own proposals approved by parent	2.73	3.00	2.36	2.58	2.87	2.80	3.00	2.85	2.63	2.50	2.71
Feedback from local											
(i) production units	2.27	2.20	2.30	2.08	1.47	2.40	1.75	2.00	1.25	2.33	1.93
(ii) marketing units	2.73	2.50	2.27	2.49	1.93	2.20	2.50	2.67	1.87	2.50	2.33
(iii) sales channels	2.36	2.60	2.09	2.25	1.43	2.20	2.50	2.17	1.62	2.00	2.02
(iv) customers	2.64	2.67	2.55	2.53	1.90	2.40	2.75	2.75	2.44	2.17	2.40
Feedback from foreign											
(i) production units	1.91	2.20	1.55	1.33	1.27	1.60	1.25	1.33	1.33	1.33	1.43
(ii) marketing units	2.09	2.50	1.55	1.78	1.67	2.00	2.00	1.67	2.33	1.83	1.84
(iii) sales channels	1.64	2.40	1.55	1.58	1.33	1.80	2.00	1.50	2.00	1.67	1.61
(iv) customers	1.91	2.80	1.91	1.97	1.77	2.00	2.00	1.92	2.67	1.80	1.98
Part of collaborative research with another enterprise	1.50	1.75	1.70	1.66	1.77	1.75	1.50	1.91	1.50	1.67	1.69
Independent local researchers	1.73	1.40	1.18	1.41	1.63	1.50	1.25	1.58	1.33	1.50	1.48

[a] Respondents were asked to grade sources on the scale, (1) never a source of ideas, (2) occasionally a source of ideas, (3) a regular source of ideas.
[b] The average derived by allocating values to the responses of (1) for 'never', (2) for 'occasionally', (3) for 'regular'.

Table 7.3(b) Sources of project ideas initiated in overseas subsidiary R&D units[a], by host country and home country

Sources of ideas: Average response[b]	Host Country					Home Country				
	USA	UK	Other Europe	Japan	Other countries	USA	UK	Other Europe	Japan	Other countries
Suggested by parent	1.93	2.06	1.81	3.00	2.11	1.94	1.86	2.10	1.50	1.67
Suggested by sister affiliates	1.66	1.73	1.59	2.00	1.78	1.65	1.57	1.74	1.25	2.33
Own proposals approved by parent	2.58	2.91	2.78	3.00	2.50	2.83	2.55	2.70	2.00	2.67
Feedback from local										
(i) production units	1.98	1.82	1.96	2.00	1.89	1.90	1.95	1.92	2.00	2.33
(ii) marketing units	2.52	2.24	2.22	2.00	1.89	2.25	2.62	2.28	2.75	2.00
(iii) sales channels	2.18	1.88	1.89	2.00	1.89	1.96	2.24	1.98	2.25	1.67
(iv) customers	2.47	2.40	2.41	2.00	2.00	2.43	2.52	2.29	2.75	2.33
Feedback from foreign										
(i) production units	1.28	1.59	1.56	1.50	1.33	1.49	1.33	1.31	1.50	2.67
(ii) marketing units	1.63	2.14	1.96	1.50	1.67	2.08	1.52	1.76	1.25	2.00
(iii) sales channels	1.46	1.88	1.70	1.50	1.22	1.79	1.52	1.44	1.25	2.00
(iv) customers	1.75	2.35	2.18	1.50	1.56	2.30	1.95	1.67	1.50	2.33
Part of collaborative research with another enterprise	1.70	1.77	1.59	1.50	1.67	1.53	1.81	1.77	1.75	2.33
Independent local researchers	1.48	1.35	1.65	2.00	1.33	1.41	1.62	1.50	1.50	1.33

[a] Respondents were asked to grade sources on the scale, (1) never a source of ideas, (2) occasionally a source of ideas, (3) a regular source of ideas.
[b] The average derived by allocating values to the responses of (1) for 'never', (2) for 'occasionally', (3) for 'regular'.

development work in support of MNE operations in third countries, this low figure for regular project idea inputs from sister R&D affiliates may indicate that the main links in formulating this work are with marketing or production in those countries. Even where the needs of overseas producing or marketing affiliates are transmitted through allied R&D units in the same country, as we have hypothesised may happen when the problems are sent to IILs, these may not be rated as project ideas, but rather as questions which the unit considers as then being articulated into projects by itself. Indeed, as shown in Table 7.3, the responding subsidiary laboratories do evaluate overseas production units, marketing, customers etc. rather more highly as sources of project ideas, though still less pervasively than might be indicated by their commitment to work for them. This indicates that, as suggested, the laboratories often link directly with other functions in these third country markets, but perceive the inputs from them as often less than fully structured project ideas.

In the light of these results it is not surprising that 77% of respondents assessed their 'own ideas approved by parent' as a regular source of project proposals. As indicated it is our interpretation that this result mainly represents the unit's formulation of research projects around inputs from other functions (production, marketing), rather than these emanating from their own creative adaptation of ideas within recent scientific output. The result does, however, indicate the degree to which overseas R&D feels itself to be subject to central scrutiny, reflecting further the extent to which the dispersion of R&D in MNEs increasingly supports a global strategic perspective, even when providing adaptation or development work.

Generally our results may be interpreted as indicating that the overseas subsidiary laboratories of MNEs have relatively little involvement with the more speculative stages of the evolution of their group's technology, i.e. their work is not science-driven in the sense of being more clearly defined by the nature of incoming technology, and the momentum of its further refinement, than by the characteristics of a determined commercial need. However, other results make it clear that these laboratories do very strongly see themselves as mediating between progress in the relevant technology and the commercial needs of the environments they support. To the extent that they may seek applications for new technology pre-emptively, i.e. without waiting for requests or problems to be transmitted to them by marketing or production units, they may then to some degree be technology-driven. Results (see Table 7.4) show[23] that 51% of respondents considered the 'rate of change of technology in the industry' to have been of major importance as a factor influencing their own growth, compared with 36% that rated 'growth of host country market' as a major factor and 22% that similarly

---

[23] See [9, Chapter 6.3] for more detailed discussion.

Table 7.4 Importance of factors influencing the growth of subsidiary R&D units[a]

	Type in involvement (average frequency[b])		
	Growth of host country market	Growth of overseas affiliate (and parent's)	Rate of change of technology in the industry
Industry			
Food, drink, tobacco	2.54	2.00	2.55
Petroleum	1.33	1.67	2.67
Metal manufactures and products	1.73	1.46	2.55
Industrial and agricultural chemicals	2.22	1.89	2.44
Pharmaceuticals and consumer chemicals	2.03	2.07	2.13
Motor vehicles and components	1.80	1.80	2.60
Industrial and farm equipment	1.75	1.25	2.50
Electronics and electrical appliances	2.29	2.00	2.39
Office equipment (incl. computers)	1.57	2.25	2.57
Other manufacturing	1.83	2.00	1.67
Total	2.05	1.91	2.39
Host country			
USA	2.34	1.68	2.42
UK	1.88	2.12	2.42
Other Europe	1.75	2.07	2.21
Japan	2.50	2.00	3.00
Other Countries	1.56	2.00	2.22
Total	2.05	1.91	2.37
Home country			
USA	1.92	2.10	2.33
UK	2.00	1.81	2.43
Other Europe	2.17	1.78	2.38
Japan	2.50	1.25	2.00
Other Countries	1.67	2.33	3.00
Total	2.05	1.91	2.37

[a] Respondents were asked to grade the influence of each factor on the scale 1: of no importance, 2: of some importance, 3: of major importance.
[b] Average derived by allocating values to the responses of 1 for 'no importance', 2 for 'some importance', 3 for 'major importance'.

rated 'growth of overseas affiliate and parent markets'. Further 'rate of change of technology' was considered a major factor influencing their growth by 58% of LILs, 48% of IILs and 39% of SLs, but by only 33% of laboratories regularly doing basic/original research. Thus those laboratories that are primarily concerned to support the ability of the group to sustain its position in global competition with distinctive product variants for dispersed markets do see their need for this role to be directly related to the rate at which relevant new technology is coming on stream. The distinctive nature of the units that are influenced by wider market growth is also indicated. Thus 33% of units regularly doing basic/original research considered 'growth of overseas affiliate and parent markets' to have been a major influence on their growth (compared with a range from 20% to 26% for other types of research), whilst the figures for laboratory types were 28% for IILs, 25% for LILs and 23% for SLs.

While our results are strongly in line with the expectations derived from the perception that overseas R&D laboratories play a variety of roles in supporting MNEs' pursuit of sustained global competitiveness, there is only limited indication of any substantial direct influence on their evolution or behaviour deriving from competitive factors. Of possible competitive influences on recent decisions with regard to the development of subsidiary R&D units (see Table 7.2), only one relating to the broad characteristics of the competitive environment ('a large and growing market where R&D is seen to play a critical role') emerged as of any relevance, with 29% assessing it as a major factor. None of the possible forms of oligopolistic reaction within technology received any notable response. A modest exception to this was discerned in the analysis of those parent laboratories which had contemplated but not implemented overseas R&D. Here 22% of respondents said that their consideration of subsidiary laboratories overseas had been provoked by 'increased internationalisation of R&D by our rivals', though only US (26% of their replies) and Japanese (31%) parent laboratories acknowledged this factor.

7.4 CONCLUSIONS

Overall our survey results provide broad support for the view that overseas R&D is playing an increasingly distinctive role in many leading MNEs. Furthermore the evidence is compatible with the expectation that the growing importance of this overseas R&D matches the emergence of general perceptions of the evolution of global perspectives on competition among these enterprises, and more specifically of the crucial role to be played by an increasingly global approach to innovation. Whilst, where relevant, it is clear that the work of MNEs' overseas subsidiary R&D laboratories may still encompass the traditional role of adapting for the local market (or to local production conditions) products already successfully innovated, our results also indicate the manner in which these facilities may now assist in the effective implementation of the global perspective on innovation.

Two aspects of the way in which the evolution of overseas R&D facilitates its role in supporting the MNE's dispersed and differentiated approach to innovation are indicated. The first of these is that development work

Internationalisation of R&D: Organisation and Motivation — 161

replaces adaptation, so that the R&D units work with marketing and/or engineering units to derive distinctive product variants within the context of the group's new technology. Secondly the survey replies indicate that some overseas R&D units provide these market-support (i.e. demand-side) services for a range of markets. This can take two forms. The first of these, in one sense mainly a geographical extension of the traditional adaptation role, is to provide 'trouble-shooting' assistance to producing or marketing units in other countries, where their own capabilities are inadequate to solve problems emerging in the pursuit of product or process adaptation. Though this is a geographical extension of the traditional role of overseas R&D, it is often carried out by different types of facilities, i.e. by IILs rather than by the mainly local-market-oriented SLs. This reflects the fact that IILs encompass both more advance scientific ability and an established expertise in international (though intra-group) communications. The second geographically extended, market-oriented, role of overseas R&D is to develop new products (for global or regional markets) to be produced by an allied WPM subsidiary. Here LILs with a development orientation predicated on strong links with marketing and production are the key type of laboratory.

Whilst our results clearly support the view of overseas R&D playing crucial roles in the global innovation programmes increasingly adopted by MNEs as part of their worldwide competitive strategy, there was less decisive support for the complementary expectation that overseas R&D facilities would also play a role in the programmes of basic/original research implemented to underwrite future innovations. This is indicated both by the relatively limited performance of basic/original research in overseas laboratories, and by the low valuation of distinctive host country scientific capabilities as an influence on these units. Against this it was found that where basic/original research was carried out in overseas facilities there was a very strong propensity for this to be integrated into internationally networked programmes. Also overseas laboratories that performed basic/original work had a well above average tendency to specialise on such work, with complementary performance of other types of R&D relatively rare. Generally this indicates that few overseas R&D facilities of MNEs incorporated the full integrated creative process, either developing basic work performed elsewhere or providing basic/original results *for* development elsewhere.

Parent laboratories' attitudes to their overseas R&D subsidiaries seem to reflect the roles these units have been revealed as playing in the MNEs' global competitive programmes. Thus relatively few of these overseas facilities are now autonomous and beyond central scrutiny. Nevertheless it also seems that relatively few are involved in playing specialised roles in programmes imposed by central laboratories. It appears that the product development role adopted by many of these units, often involving intermediation between centrally-generated new technological possibilities and the needs of distinctive markets, frequently endows them with a sort of 'supervised-freedom'. In this approach the parent laboratory seeks to ensure itself that such overseas units have access to, and make the best use of, the group's most radical technology, but in ways that involve much of their own distinctive creativity implemented in order to maximise group effectiveness in the full range of differentiated national, or regional, market environments.

7.5 REFERENCES

[1] C.A. Bartlett and S. Ghoshal (1990), Managing Innovation in the Transnational Corporation, in C.A. Bartlett, Y. Doz and G. Hedlund (eds), *Managing the Global Firm*, Routledge London, pp. 215–255.

[2] Bowker (1987), *Directory of American Research and Technology*, 22nd edition, R.R. Bowker, New York.

[3] J.H. Dunning and R.D. Pearce (1988), *The World's Largest Industrial Enterprises 1962–1983*, Gower, Aldershot.

[4] Longman (1987), *Industrial Research in the United Kingdom: A Guide to Organisations and Programmes*, 12th edition, Longman, Harlow, Essex.

[5] Longman (1988a), *European Research Centres: A Directory of Scientific, Technological, Agricultural and Medical Laboratories*, 7th edition, Longman, Harlow, Essex.

[6] Longman (1988b), *Pacific Research Centres: A Directory of Organisations in Science Technology, Agriculture and Medicine*, 2nd edition, Longman, Harlow, Essex.

[7] R.D. Pearce (1989), *The Internationalisation of Research and Development by Multinational Enterprises*, Macmillan, London.

[8] R.D. Pearce (1988/89), World Product Mandates and MNE Specialisation, University of Reading, Department of Economics, Discussion Papers in International Investment and Business Studies, Series B, Vol. I, No. 121.

[9] R.D. Pearce and S. Singh (1991), *Globalising Research and Development*, Macmillan, London.

[10] R. Vernon (1979), The Product Cycle Hypothesis in a New International Environment, *Oxford Bulletin of Economics and Statistics*, **41**, 4, 25–27.

[11] R. Vernon (1966), International Investment and International Trade in the Product Cycle, *Quarterly Journal of Economics*, **88**, 190–207.

Chapter 8

Management of International R&D Operations

ARNOUD DE MEYER

International R&D operations, though not yet a very common activity for the international firm, have been increasing over the last decade. Why is it that international firms decide to accept the increased communication difficulties of geographically decentralized R&D operations, in order to create networks of laboratories? In this chapter we discuss the three traditional explanations of proximity to cheap labor markets for engineers, proximity to markets, and proximity to sources of technology. But on the basis of 14 case studies of large European, Japanese and North American firms, we conclude that none of these explanations can completely explain the companies' behavior. The underlying explanation seems to be the learning process that is going on in R&D. We further explore the consequences of this learning objective for planning and control procedures, the networking of laboratories, and the communication between laboratories.

8.1 INTRODUCTION

In the latter half of the 1980s the globalization of companies has become a widely studied issue in management. As part of that globalization we have observed an increased interest in global technology development. Research and Development operations, which many companies tended to carry out at home in the headquarters, have increasingly become subject to internationalization. A survey carried out by three large business periodicals in the Federal Republic of Germany, Japan and the United States [4] suggests that

---

Technology Management and International Business: Internationalization of R&D and Technology.
Edited by O. Granstrand, L. Håkanson and S. Sjölander. © 1992 John Wiley & Sons Ltd

German companies carry out 85% of their R&D at home, the US companies 91.8% and for the Japanese companies in the sample the percentage of domestic R&D is as high as 98.1%. Though one has to take these figures with the necessary grain of salt, because they are probably somewhat biased by the sample of responding companies, they indicate that international R&D is not very widespread. However, the trend is not to be mistaken. International research is on the increase. A recent survey of the National Science Foundation[1] arrives at the very similar percentage of 91.9% of domestic R&D spending for US companies, and shows that R&D spending abroad by US companies soared by 33% in 1986 and 1987, while spending at home went up by only 6%. In a study of 20 Swedish multinationals [11], it was found that the share of foreign R&D as a percentage of total R&D expenditures had risen from 20.6% in 1920 to 22.8% in 1987. In spite of the rapid escalation of domestic expenditures from 1980 to 1987 (214% in current prices), these Swedish multinationals had increased their expenditures in foreign locations even more (252%).

Data on a sample of seven large German multinationals show that the share of foreign R&D personnel rose from 13% in 1974 to 19% in 1983, while the share of foreign employment increased from 31 to 38% [21]. For Japan, the trends are less well documented, but the scarce indications available suggest that similar trends exist in this region.

Why is it that companies decide to increase the geographical decentralization of their R&D? How does this influence strategy formulation, and how is this international network of R&D operations managed? These are the questions we intend to address in this chapter. To provide partial answers, we will use the insights we obtained through case studies carried out in 14 large European, North American and Japanese companies.

8.2 THE RESEARCH BASIS

In 14 large multinationally operating companies, data were gathered through interviewing research managers and their subordinates. In some cases the users of the results of research and development work, such as product managers and/or production managers were interviewed as well. The interviews ranged from several hours to several days. Though the basis for the interviews was a checklist of items to be discussed, in most interviews the questions and answers were very open, and adapted to the specific technology and market presence of the company involved[2].

---

[1] As reported in the International Herald Tribune, February 27, 1989, Exporting R&D Operations could Hurt U.S. Economy.
[2] Two pedagogical case studies based on the research cases were published in Davidson and de la Torre [5].

Management of International R&D

The 14 companies do not form a sample which is representative of a specific industry. Three companies belong to the electronics industry, one to the food industry, three to the pharmaceutical industry, five to the chemicals industry and two to the automobile industry. Nine of them are European, two are Japanese, and three are North American companies which have been established for a long time in Europe. All, except two of them, had a long experience of carrying out Research and Development on an international scale. The number of laboratories ranged from 2 to 17. In each of the cases, the companies had several hundreds of people working in research and development. Each of the companies is considered to be successful to very successful by evaluations in the professional business press, and by judging from its financial performance over the last five years. This does not guarantee, however, the quality and success of their R&D performance. In fact in more than one company we were granted interviewing time, because managers were concerned about the performance of their international R&D and they hoped to obtain a sort of benchmark evaluation by the interviewing process.

The activities of the laboratories ranged widely on a scale from fundamental research to a high degree of applied technical customer service.

Table 8.1 Description of the sample

Company activity	Number of laboratories	Total number of employees	Range of lab staff
Automotive	2 (+ 1 test ground)	3500	2100 & 1400
Automotive	3	7800	70–7500
Electronics	6	3100	3–2500
Electronics	8 (2 more planned)	300	15–70
Food	1 central research lab.	450	id.
	18 local development centers	Not disclosed	20–250
Petrochemicals	3 (1 closed after study)	± 400	200
Petrochemicals	2 central research labs	3000	2300 & 700
	60 business labs	2000	10/15–several hundreds
Petrochemicals	8 group central labs	4560	150–1600
	7 operating labs	2280	60–1680
Pharmaceuticals	3	750	30–500
Pharmaceuticals	9	Not available	Not available
Speciality chemicals	1 primary research lab.	660	660
	13 development labs	550	40–100
Speciality chemicals	3	± 1200	Europe: 130 Japan: 70 US: 1000
Telecommunication	13	1300	40–250

However, we discarded from the analysis simple process engineering activities. Even in the definition of what research and development are, we found an enormous variety. What was classified by one company (even in the same industry) as research, would probably be named applied development by another company. It did not seem very useful to provide a table with the number of research and development laboratories according to their technical activities at this stage. Whenever necessary in the analysis we will make the distinction between development and research.

8.3 WHY INTERNATIONALIZE R&D? —THE TRADITIONAL REASONS

The disadvantages of international R&D operations can be summarized in one concept: increased difficulties in communication. This in itself leads to difficulties in coordination, reduces the speed of the development and can decrease the productivity of the R&D process. Through spreading out the resources in R&D, the company also runs the risk of ending up with laboratories which are below the critical mass. Thus there are more than enough reasons to refrain from internationalizing R&D. What is then the positive element that makes companies decide to accept these disadvantages and go ahead with the creation of laboratories abroad?

Traditionally, the management literature [20] has indicated three categories of reasons: increased access to cheap R&D labor, proximity to the market and the customers, and access to the best sources of technological development. Each of these three reasons finds some support in the traditional theories on the internationalization of the firm.

According to neo-classical economic theory, companies internationalize their operations (in particular their production organizations) to take advantage of cost differentials in different countries. A laboratory can be considered to be a production site of know-how. Since the cost of a laboratory is to a large extent determined by the cost of professional scientists and engineers, one would expect companies to move to countries where scientists are inexpensive. With a few exceptions, this trigger for internationalization of R&D is weak, because one has to be fairly confident that the cost differential will exist over a long time. Indeed the creation of an R&D laboratory does not happen overnight. To be able to prove itself the laboratory has to build up a base of know-how, and have the time to produce its first results. Our case studies indicated that this requires 5 to 10 years, and few locations can claim to have cost differentials for professional personnel that will extend over such a long period of time.

A more sophisticated form of this argument is used in the case of multinational companies with headquarters in small countries. Their demand for professionals sometimes exceeds the total supply of the country. It is said for example that Philips in the Netherlands requires more engineers per year than the universities produce. The same is true for Swedish companies [10].

The data available on international R&D do not confirm such a movement triggered by cost differentials or availability of engineers. On the contrary, previous studies [18,3,10,12] and also my own case studies, suggest that the huge majority of R&D investments abroad tend to be in the free market-oriented, industrialized, high labor cost countries. Consequently, this argument based on cost advantages does not seem to be the primary explanatory factor in the internationalization of R&D. Having said this, the availability of engineers was often mentioned in the interviews as a secondary consideration in the choice of location of the laboratory site.

The second category of reasons has to do with the market and is an extension of earlier theories on international product life cycles. Multinational companies have to become more responsive to their markets, have to adapt products to the national markets, and cannot always rely on global marketing and global products to be successful. The best way of doing this is by being present in this market and by working closely together with users and customers. This requires the local presence of at least one technical support laboratory. These technical support laboratories seem to have the almost unstoppable tendency to grow into local product development centers [18]. Thus one gets a proliferation of market-oriented product development laboratories in the larger markets.

The third approach is the one usually preferred by technologists. Laboratories abroad are created to tap local sources of scientific and technological development. Technology development is very parochial [1,12], and as a consequence one finds an uneven distribution of technology development across the world. Consequently, there are concentrations of sources of technological development, such as the Californian academic network for genetic engineering and the tight network of automotive suppliers in Japan and central Western Europe. To gain access to the output of these concentrated sources of technology, one has to be physically close to them, and participate in the local network of technology production.

Transaction cost approaches to market structures (see for example [19]) support this category of triggers for internationalization. Given the difficulties of trading in technological know-how, in particular the fact that one often needs to transfer people in order to transfer technological know-how, the internal mechanism of creating a laboratory abroad close to the source of know-how can be more efficient than a pure market transaction through a contract, i.e. buying technology.

8.4 VALIDATION OF THESE MODELS

These three categories of reasons are often used to explain the result of the trade-off between the advantages of internationalizing R&D and the disadvantages of increasingly difficult communication and the loss of scale economies in R&D. None of them, however, could completely explain the reasons for internationalization as they were given to me in the case studies. If one checks the results of the interviews with these three categories, all of the 14 companies and most of their foreign laboratories can be classified according to three categories. Consequently, this taxonomy in three groups does not seem to be very helpful.

One R&D manager of a European electronics firm phrased his answer this way:

> The key element of success in development is to be able to rely on good skills. We are located in [a small city far from larger cities]. We were created here. But it will be obvious that it is nearly impossible to attract everyone who is needed for good development in one location. Certainly this is not an ideal location for it. It is a very rural area, which might have a special kind of attraction for those who do not like city life, but that is not sufficient to attract all engineers or scientists who are needed. Moreover, successful development requires a constant exchange of ideas, not only inside, but also outside the firm. In [this small city] there are virtually no other firms than us who are active in similar fields. There is, except for a quite powerful university, no opportunity for an extensive exchange of ideas.

A superficial analysis of this quote would lead us to think that lack of a qualified workforce had led this firm to expand its laboratories internationally. But the enthusiasm for the exchange of ideas, the urge to belong to a network of information exchange and discussion was much stronger than the need to have access to pools of engineers. The same desire to belong to such a network was expressed by several other managers. But let us not be mistaken. Although the word network was used by virtually all of the interviewed managers, their view of a network was not limited to a technical network of research institutions, laboratories, universities, etc. In the description of these networks, suppliers, subcontractors, customers, users, government representatives and competitors were explicitly mentioned.

Why do companies want to belong to these networks? Here we come to the essence of what in our opinion drives internationalization of the R&D function. The core concept in each of our discussions was 'technical learning'.

The network that is created by building up an international R&D operation should help the company to develop better early warning systems, shorter routes to commercialization and stronger market orientation in product development. At the same time, the company will have to rely to a far greater extent on and compete for external sources of technological and market knowledge.

8.5 TECHNICAL LEARNING AS A FACTOR OF INTERNATIONALISING R&D

Research and development have a role to play in innovation. This role is twofold: solving technical challenges which are part of the innovation processes, and contributing technology-based ideas to the concept definition, but also creating, maintaining and expanding a knowledge base which will enable future innovations. Learning as part of the R&D process has recently been stressed by several scholars [13,14,15,16]. This learning aspect becomes even more dominant in the case of international R&D. If R&D's task were to be limited to technical problem-solving, the need to geographically decentralize R&D operations would probably be far less urgent. Access to problem-solving capacity, namely the engineers, may still be a reason. But though this access is sometimes a secondary trigger for internationalization, it is rarely dominant in the choice of the number of foreign R&D sites and their location.

Technical learning in new product and process development is not limited to internationalization, and internationalization is not the *conditio sine qua non* of learning. But learning about different markets, different problem-solving methods, different sources of technological progress, different cultures, different competitors, and rapid diffusion of that learning throughout the organization, is definitely enhanced by creating an international network of R&D laboratories. In other words, apart from the result-oriented problem solving, the R&D group has to learn for the company, to enable the company to pursue transnational strategies in the future, and effective technical learning requires the R&D group to go international.

Learning can be defined as the process within the organization by which knowledge about action-outcome relationships and the effects of the environment on these relationship are developed [9]. Unlike individual learning which involves relatively permanent changes in an individual's behavior, technical learning by an organization involves the development of a know-how base owned by an organization, which would make such change possible. The outcome of the learning process is knowledge that is distributed across the organization, is communicable among members, has consensual validity, and is integrated into the working procedures of the organization.

If one applies this to the technical learning process, it appears that exposure to sources of knowledge in different countries is important, but that in order to be effective, one has to create mechanisms on an international scale to diffuse, validate and integrate the new knowledge across the whole network of laboratories. Validation of the information has often to do with the *credibility* of the laboratory *vis-à-vis* its partners. To produce unique information, i.e. information which could not have been produced in the central R&D laboratory, the foreign laboratory must be able to take advantage of the local circumstances. A certain degree of diversity among the laboratories is thus needed. Diffusion, validation and integration are heavily determined by the quality of the formal and informal communication system. The informal communication system is highly dependent on the organizational network which exists in the firm.

Furthermore, from a management point of view, we are interested in learning systems which can be managed, where conscious efforts by research managers can improve the effectiveness of the international R&D network. The creation of laboratories in locations far away from the headquarters indicates that the firm wants to learn through an organization, and not through an individual. If learning were to be dependent on a single individual, the firm would probably be better off by paying the individual enough so that he would move to the central R&D laboratory, or by building a monitoring device around that person. Thus the technical learning must become part of the procedures. The focal point of these procedures is often the planning process.

Consequently, three categories of managerial action deserve our attention in the analysis of the case studies. Firstly, how can the planning and control process, which is traditionally the formal link between strategy and R&D, contribute to the technical learning process? Secondly, networking seems to be the dominant form of integrating the business perspective and the technological perspective. And thirdly, technical learning requires diffusion of technical knowledge. Effective diffusion of knowledge requires communication. Geographical dispersion forces companies to take a different perspective on their communication networks in R&D. Each of these three themes will be discussed in the following paragraphs.

8.6 THE PLANNING AND CONTROL CONTRIBUTION TO TECHNICAL LEARNING

Behrman and Fischer [3] analyzed in depth how planning and control are exercised in international R&D. Following a taxonomy of managerial styles proposed by De Bodinat [6], they assign the managerial styles of the companies in their sample to four discrete points on a continuum from absolute centralization of the decision-making over participative centralization and supervised freedom to total freedom. We have found in some of our case studies a similar result [8]. However the categorization in discrete points on this continuum is artificial, and we have asserted that planning and control, and managerial style of decision-making can move on the continuum over time, depending on the technological characteristics of the specific foreign laboratory involved. The position on this continuum is dynamic rather than static, and does not depend in the first place on the character-

Management of International R&D

istics of the parent company, but rather on the characteristics of the individual laboratory. The parent company's culture plays a role here but this role is rather of the second order.

But a more interesting observation is that, in each of the companies that we did not consider to be unsuccessful, planning and related activities had a strong learning component. Why do we say 'not unsuccessful' companies? Due to our research methodology we have no hard measure of success. However, in two of the companies we studied, the planning procedure was changed shortly after our interviews. The mere fact that the company reorganized its procedure can be judged to be an indication of below average performance of the planning process before the reorganization. There is of course no indication that companies that did not reorganize are performing well.

How did this learning component find its expression? In many of our cases the managerial decision-making on planning issues was closer to participative centralization than to supervised freedom. However, an enormous effort was required to involve local laboratory personnel and local laboratory management in the formulation of the objectives, and the means to reach those objectives. One company had a two-year planning cycle with internal scientific conferences which were used to determine research targets but also to exchange information between strategic planning, the business units and the scientists. And in one of the specialty chemicals companies the research manager said:

> Tied to the planning process are meetings of the heads of R&D centers once a year in [headquarters' location] for what may be called information meetings. These meetings do not bring all centers together at once. In fact there are three meetings each year for the heads of the Southern hemisphere, South East Asia and the rest of the world. These meetings are set up to exchange information about the latest field results and market developments around the world. Following these meetings there is a period of information digestion. During this period the [staff of the central R&D] are involved in extensive travelling to various centers to discuss issues brought up during the information meetings. The final planning meeting takes place two months after the information meetings.

These and other remarks at this company showed that the whole planning cycle was used more to educate each other and diffuse information than strictly for planning. A one-year budget cycle was even said to make no sense in this company, since development cycles were from five years upwards. But the educational value to the company of the planning process was very important. The same company also has a central database for general access. All data generated by the different development centers are entered in the database in a uniform format. The database, which is accessible to all researchers and product managers, is considered to be an excellent means of diffusion of information as well as control of the ongoing activities.

In one of the other companies, which was characterized by its own management as an example of participative centralization, the central research staff is responsible for the yearly formulation of defining and assigning R&D projects to the different centers. But since this central staff is very small compared to the total number of laboratories and professionals, they have to rely entirely on discussions with the operating units to gain an insight into technology, markets and process capabilities. The travelling schedule of the staff of this central R&D body was very impressive, and clearly an indication of a great deal of personal communication. The control procedures of this company, mainly a set of standardized progress reports, were equally used as educational instruments and diffused very quickly and efficiently to well chosen targets in R&D and marketing. Again, the whole planning and control procedure included a strong educational element.

In one of the companies a major tool of planning was what they called the VIP's or Very Important Projects. They were considered to be a centerpiece of the research efforts, since every laboratory was supposed to contribute to it. Such a project had of course a strong integrating influence, but it was also a way of exchanging ideas between different laboratories. Again we find planning as a learning tool! In the same company a number of advisory and strategy boards were involved in the planning process. Though the central R&D group was instrumental in initiating the board meetings and coordinating the whole process, laboratory managers were heavily committed throughout the process. A core element of the planning process was the communication and cooperation in the various meetings. The board meetings, in which researchers, representatives of the business units and distinguished outsiders participate, were considered to be unique opportunities for cross-boundary contacts. Laboratory managers spend about two and a half to five days per month in coordinating and planning meetings, which are organized at the different sites to 'give the researchers the opportunity to meet regularly the members of the central R&D team'. Planning and evaluation of ongoing projects were really used as an organizational learning experience for the company.

What was striking was that in the two less successful companies the educational element of the planning process was conspicuously lacking. In one case the planning process was carried out by a Strategic R&D Planning group which only at the highest level interacted with the Strategic Business Planning Group. Perhaps, in the end, the result of the planning exercise was a fair reflection of technological potential and market opportunities and that in that sense the process was satisfactory, but there was clearly no attempt to use the planning process as an exchange of ideas and a process of mutual education. In the second case the planning at top level was made by a board composed of the technological and business managers, but at a lower level there was no other interaction than the one needed for short-term problem-solving. Also between different groups in R&D there was little interaction. Synergies between different R&D groups were deemed to be almost nonexistent, and creating them was considered a waste of time. A comment made by a business manager of a small diversification project in this company was quite revealing: 'only when we [the business managers and the technologists of different R&D sites] started educating each other about what we knew, we started making some progress'.

In order to gain the benefit from internationalization of R&D in the organizational learning process of the company it seems that the planning, control and evaluation process, independent of its positioning on the scale from absolute control to absolute freedom, has to be transformed into a learning process. It is interesting to note that Meyers and Wilemon [16] come to the conclusion, on the basis of an exploratory survey of learning in new product development teams, that the strongest inhibitors for learning have to do with establishing and maintaining clear project objectives. This is clearly an element of the planning process. De Geus [7] gives a description of Shell's planning process and asserts that planning should be a learning process, and that one of the objectives for improving planning systems should be the acceleration of the learning process. Consequently, common scenario building, common preparation of the strategy formulation and projects common to different laboratories can contribute to improved linking of R&D and strategy through improved learning. In the tradeoff between planning efficiency and learning of the organization, the choices made should favor the learning side.

8.7 NETWORKING AS A CORE ELEMENT OF THE ORGANISATION

The rapid diffusion of data, information and knowledge is part of the learning process and it also serves to accelerate the learning. One aspect of this diffusion has to do with communication in an environment where the difficulties of communication are compounded by geographical distances. A second element is the networking. Let us look at this second element first.

In all of our case studies the concept of networks was raised at some point. Networks were considered to be the new and most appropriate way of organizing the relations within an international R&D operation. Scholars of international business have also shown considerable interest in the concept in recent times [10,2,14].

Thinking about networks in communication terms, one has to keep four elements in mind: the roles of the nodes, the density of the communication on the links, the ties to other networks, and the dynamics of node roles and link density [17].

The role of the nodes becomes in this context what the specific role is of the laboratory in the learning of the organization. What is its mission or charter? The R&D manager of one of the electronics companies insisted very strongly on the fact that to have a successful contribution of the foreign laboratory to the firm, the laboratory needed a worldwide charter, i.e. a worldwide responsibility in the problem-solving and learning process. But charters must be dynamic and be adapted to the changing characteristics of the information sources among which the foreign laboratory is implanted. In one of the firms in the cases where the senior R&D management and customers were not satisfied with current performance, part of the lack of satisfaction could be explained by a lack of a clear charter for each of the laboratories in the learning process. The case was interesting in the sense that it was one of the cases which had been described at length in earlier studies of international R&D management, and the present situation could be compared with an account written in the middle of the 1970s. First of all, the management felt uneasy about being asked about the role of the different laboratories and fell back on some of the descriptions which had existed in the early 1970s. They admitted, however, that these role descriptions were more history than reality. In one particular case the charter consisted, among other things, of using the process capabilities of a particular factory to develop worldwide product applications. Originally, factory and laboratory were close to each other. But an incremental change in laboratory location and factory know-how had gradually rendered the charter of the laboratory obsolete. However, formally the charter had remained the same. To conclude, one can say that in order to contribute to the learning process of the company, the different laboratories must have a clearly and dynamically defined charter which is known and accepted worldwide.

The second characteristic of the network is the local external network. This is the main mechanism through which the local laboratory can fulfill its role of local learning. The density and quality of the communication with local partners are a measure of the laboratory's effectiveness in tapping the local network.

The information and knowledge learned locally have to be diffused in the company. A self-organizing local external network can only be effective in terms of innovation if it is linked to a strong internal network [2]. This is partially an issue of intra-company communication. But it is also related to how well the laboratory is embedded in the network inside the company but outside R&D. In one company the senior management had taken two successful actions to strengthen the intra-company network. First they had chosen persons with a high visibility in the headquarters to act as laboratory directors or as senior researchers in the laboratory. These people had the explicit role of being the ambassadors of the laboratory at the headquarters and the other functions in the company. Second, realizing that there was a gap between decentralized laboratories and centralized marketing, some of the marketing functions were decentralized to the same locations as the laboratories. One sees here two complementary actions: a strengthening of the local intra-company network (decentralizing marketing) and a strengthening of the international intra-company network (the ambassador role).

This ambassador role emerged in several of the case studies. Whatever the name, time and again attempts were made to increase the visibility of the foreign laboratory in headquarters or in important subsidiaries, by using the personal credibility and visibility of the so-called ambassador to represent the laboratory. Other actions that were described to us to strengthen the intra-company network had to do with intra-company scientific conferences, tours of selected scientists and engineers through the different laboratories, projects cutting through the different laboratories, worldwide task forces, etc.

Apart from the local external and the intra-company networks, we were several times confronted with the existence and importance of the external international network. An analogy might help us to understand this. The European electrical grid is widely interconnected, and there are constant flows of electricity from one country to another. To a country like Belgium for example it is of the utmost importance to be connected to the French and the German grids, so that it can rely on both to avoid loss of tension or frequency. However, what is also important is that, in case the line to France for some reason is cut off and Germany has no spare capacity to support the Belgian grid, Germany is interconnected with France and can channel power from France through its own network to Belgium. The example is far removed from the organizations we study here, but the comparison can nevertheless help us to understand a characteristic of these organizations. The external local networks of the different laboratories are also interconnected, and careful management of this international network external to the company can provide a boost to the effectiveness of the learning and diffusion process in the company. In the most extreme case the external network can be a more efficient diffusion mechanism than the internal communication network. This does not necessarily have to be the result of a breakdown of the internal networks. It can simply be that two laboratories work on very different projects, but with a common technology provided by an external supplier. The synergy between the two projects might be obscured by different perceptions about the two projects within the company while the supplier might be able to provide some linkages between the engineers working with a similar technology in two different laboratories.

We observed several examples of how these external networks can be useful in the management of R&D. Users of micro-electronic components find subsidiaries of the major suppliers such as NEC or Motorola in each of the countries where they have laboratories, and we observed that communication between the sales subsidiaries of these suppliers exists. There is usually also a fairly strong network between the academics with whom the company has contacts in the local R&D laboratories; they meet each other at the same professional and scientific conferences, read each others' papers, and exchange data on the research projects they might carry out for the same company, but in different countries. In one of the case studies, the R&D management had consciously attempted to nurture these international links between academics by organizing so-called private conferences to which all the academics who worked for the company were invited to make presentations. The R&D management of this company told us explicitly that the main goal of these conferences was to create an invisible network of academics working for their company.

8.8 COMMUNICATION

Communication in R&D is probably one of the most widely researched topics in R&D management, and the most consistent result out of all this research seems to be that personal contacts are the best form of communication in R&D. The seminal work of Allen [1] has demonstrated this amply, but it has been confirmed by many other studies in the USA and Europe. It does not seem necessary to add here more examples of general communication problems and solutions in R&D. The most interesting aspect of this communication problem in an international context is the added difficulty of geographical distances. Indeed it makes the core element of communication in R&D, the informal personal contact, much more difficult.

Obvious solutions to overcome this had to do with the traditional methods of integration (common project teams, exchange of researchers, etc.). In all but one of the companies we studied, the travel budget ranged between 5 and 7% of the total R&D budget, and this was generally not considered to be too high. Though there were travel restrictions, they had more to do with the disruptive effect of excessive travel on the work at home (in the local network), than with the direct cost of it. However, everybody recognized the burden that international travel brought to scientists' time utilization. Two types of actions were considered to cope with this. Firstly, an emphasis on documentation of the work. Documenting research results is important in any situation, but the discipline required as well as the effort to disseminate the resulting reports appeared to be of a different order of magnitude than in a normal single site laboratory. Secondly, many expectations were built up about the possibilities of electronic communication. Consequently, all of the companies we studied were experimenting with more

Management of International R&D

or less sophisticated forms of electronic communication. These experiments ranged from the first careful attempts to use electronic mail to full-fledged video-conferencing systems.

One of the main questions raised about these electronic communication systems was the extent to which they could replace direct personal contacts. Confidence did not seem to be high that electronic communication, even in its most sophisticated form, such as in video-conferencing, could be more than a temporary replacement to the direct face-to-face contact. 'Video-conferencing, integrated CAD/CAM databases, electronic mail and intensive jet travel all contribute to lower the communication barriers. All things considered, however, the most effective form of communication, especially in the beginning, is a handshake across a table to build mutual trust and confidence. Then and only then the electronics can be really effective,' was the way the senior product development manager of the company that had deployed the comparatively most sophisticated electronic communication systems in our sample, described it. In each of the companies that had built up some experience of electronic communication systems, we could sense the same idea as is expressed in this quote. Electronic communication can make a very valuable contribution on condition that a certain level of confidence between partners already exists. The 'handshake' is an important prior condition for the effective use of electronic communication systems.

This level of confidence can seemingly only be built through personal face-to-face contact. It has a tendency to drop over time, even if there is intensive use of sophisticated electronic systems. One of the engineers with a considerable experience of video-conferencing summarized his own attitude by saying:

> [Although it is a great system], I have two difficulties with it. I still cannot express emotions on a video-conferencing system. It seems so silly to become angry, to joke, to deviate from the subject and to talk about your family, to complain about your boss, all those things you need to do to get to know each other. And I am never sure that my colleagues at the other end are not taping me, to use my own words against me. I know it is silly, because I am not scared of taping at the phone, but video-conferencing meetings still create much more official commitments than a simple phone call.

Thus even with the best electronic communication systems confidence between the members of a project team spread out over the globe seems to decay like nuclear radiation over time, even if they have real time contacts through electronic mail and billboards, computer conferences, video-conferencing systems and the telephone. Confidence between engineers has perhaps, like nuclear radiation, a half-life time. Thus regular face-to-face contact still seems necessary to boost that confidence to a level high enough for efficient team work. We have called this decay of confidence and the regular face-to-face contacts needed to restore the confidence level to its initial level the half-life time effect of electronic communications in international R&D.

A second issue which was regularly raised in the context of international communication was the increase in problems about security. Though it seems to be a mere technical problem to protect databases and communication lines from intrusion by outsiders, in several companies, the risk of access by third parties inhibited a full commitment to computer communication in R&D. Consequently, some of the companies we studied were reluctant to use international computer networks for more than purely routine results. More sensitive materials, especially discussions about the most recent technical developments, which are so essential to learning in a company, were still kept out of the realm of electronic communication systems.

8.9 CONCLUSION

The thesis we have developed in this chapter is that, though the classical theories on the internationalization of the firm explain some of the reasons why companies develop international R&D operations, the underlying explanation can be summarized in the concept of learning. The contribution of international R&D to the technological strategy of the firm is in the improvement of the company's learning about the long-term evolution of markets, technologies, competitors and suppliers. If one accepts that point of view, the effectiveness of the international R&D operations will be determined by the company's ability to manage that learning process. Linking the learning in R&D to the technological strategy takes place in most cases through the planning, evaluation and control processes, and we assert that our planning processes should be reviewed with this learning as an objective in mind. Planning should be a learning exercise, not in the planning process, but in the technological and market evolutions. Linking the learning to the technological strategy also requires an extremely well organized diffusion of the knowledge throughout the firm. This diffusion can be stimulated by paying special attention to communication, but also by seeing the R&D organization as a network of laboratories which are connected with each other inside as well as outside the company.

8.10 REFERENCES

[1] T.J. Allen (1977), *Managing the Flow of Technology*, M.I.T. Press, Cambridge, Mass.
[2] C.A. Bartlett and S. Ghoshal (1989), *Managing across Borders, The Transnational Solution*, Harvard Business School Press, Boston, Mass.
[3] J.N. Behrman and W.A. Fischer (1980), *Overseas Activities of Transnational Companies*, Oelgeschlager, Gunn & Hain, Cambridge, Mass.

[4] K. Brockhoff, T.G.J. von Ghyczy, and W. Wilhelm (1988), Die Grossen Drei im Test, *Manager Magazin*, October, 185–197.
[5] W.J. Davidson and J. de la Torre (1989), *Managing the Global Corporation*, McGraw-Hill, New York.
[6] H. De Bodinat (1976), *Influence in the Multinational Entreprise: The Case of Manufacturing*, unpublished doctoral dissertation, Boston, Harvard Business School.
[7] A.P. de Geus (1988), Planning as Learning, *Harvard Business Review*, **66**, No. 2, 70–74.
[8] A. De Meyer and A. Mizushima (1989), Global R&D Management, *R&D Management*, **19**, No.2, 135–146.
[9] R.B. Duncan and A. Weiss (1979), Organisational Learning; Implications for Organisation Design, in Staw, Barry (ed.), *Research in Organisational Behavior*, JAI Press, Greenwich, pp. 75–123.
[10] L. Håkanson and U. Zander (1986), *Managing International R&D*, Stockholm School of Economics, Stockholm.
[11] L. Håkanson and R. Nobel (1989), *Overseas Research and Development in Swedish Multinationals*, paper presented at Academy of International Business Meeting, Singapore, December.
[12] J.M. Harris (1987), The Global Management of R&D Resources, *Outlook*. Booz, Allen & Hamilton, New York, NY.
[13] K. Imai, I. Nonaka and H. Takeuchi (1985), Managing the New Product Development, in Clark K.B. and Hayes R.H., *The Uneasy Alliance*, Harvard Business School Press.
[14] K. Imai and Y. Baba (1989), *Systemic Innovation and Cross-Border Networks*, Discussion Paper No. 135, Institute of Business Research, Hitotsubashi University, Kunitachi, Tokyo.
[15] M.A. Maidique and B.J. Zirger (1985), The New Product Learning Cycle, *Research Policy*, **14**, 299–313.
[16] P.W. Meyers and D. Wilemon (1989), Learning in New Technology Development Teams, *Journal of Product Innovation Management*, **6**, 79–88.
[17] N. Nohria and S. Ghoshal (1989), *Requisite Complexity: Organising Headquarters-Subsidiary Relations in MNCs*, Insead Working Paper No. 90/74, Fontainebleau.
[18] R. Ronstadt (1977), *Research and Development Abroad by U.S. Multinationals*, Praeger, New York, NY.
[19] D. Teece (1981), The Multinational Enterprise: Market Failure and Market Power Considerations, *Sloan Management Review*, **22**, No.3, 3–17.
[20] V. Terpstra (1977), International Product Policy, The Role of Foreign R&D, *Columbia Journal of World Business*, Winter, 24–32.
[21] M. Wortmann (1990), Multinationals and the Internationalisation of R&D: New Developments in German Companies, *Research Policy*, **19**, No. 2, 175–184.

Chapter 9

Internationalization and Diversification of Multi-technology Corportions*

OVE GRANSTRAND AND SÖREN SJÖLANDER

This chapter explores the nature of and relations between internationalization and diversification in a sample of Japanese, US and Swedish firms. The firms are typically large, multinational and multi-technology, that is they operate in several nations and technologies. The fundamental importance of technology diversification is demonstrated. A tentative causal model of technological progress, diversification and international competition is presented. It is argued that technology-based product diversification is complementary to internationalization at firm level. Reaping economies of scale and scope across nations, products and technologies present new challenges to technology management in international business.

---

*This chapter was written as a paper within the research program 'Economics and Management of Technology' at the Department of Industrial Management and Economics at Chalmers University of Technology. The research behind this chapter has been carried out in collaboration with Åsa Lindholm, Christer Oskarsson and Niklas Sjöberg. The research has been financed by the National Swedish Board for Technical Development, the Salén Foundation, the Wallander Foundation, the Scandinavian Sasakawa Foundation and Chalmers University of Technology. The research has been carried out under the auspices of the Institute for Management of Innovation and Technology (IMIT).

Technology Management and International Business: Internationalization of R&D and Technology.
Edited by O. Granstrand, L. Håkanson and S. Sjölander. © 1992 John Wiley & Sons Ltd

9.1 INTRODUCTION

9.1.1 Purpose of this Chapter

This chapter examines the two distinct phenomena of internationalization and diversification of technology in large corporations and some of their causes and possible effects. Important distinctions made in this chapter are between market diversification, product diversification and technology diversification, referring respectively to the expansion level and process of a firm's range of markets, products and technologies in which it operates. Market diversification in turn may refer to input (factor) markets or output (product) markets. Broadly speaking, internationalization is the special case of diversification into various national markets. Along with the diversification distinction, one may distinguish between multinational, multi-product and multi-technology firms. The latter concept is new and refers to firms that operate in several (at least three) different generic technologies (see [11] for a more precise definition).

9.1.2 Background

In a Coasian sense diversification of a firm, be it market, product or technology diversification, substitutes for the market. Cost and revenue considerations, be they transactional, scale or scope oriented, might be used in explaining differences in internationalization and diversification behavior among firms. However, an important factor behind relative costs and revenues in comparing modes of market organization is management learning, especially learning how to handle unique competence assets, including a portfolio of technological competences. With this view, the evolution and transfer of technology management skills and practices become a central explanatory variable. For example, differentials in international management learning and technology management learning may then explain differentials in the emergence of multinational corporations (MNCs) and multi-technology corporations (MTCs), other things being equal. Altogether, analysis of diversification patterns and economies of scope might teach us something about the ever changing scope of the industrial firm.

9.1.3 The Empirical Study

The general purpose of the project behind this chapter is to compare the best management practices regarding R&D, technology and innovation in Japan, the USA and Europe, and to try to assess the role of technology management in international competition in a comparative management and economic perspective. For this purpose, interview series and surveys in companies in the three areas were carried out. This chapter is based on 91 interviews with technology executives among 12 Swedish, 14 Japanese and 16 US corporations and on questionnaire surveys on the corporate and product area level, respectively, among these firms. All of the firms are technology-based, multinational and among the largest in each country. It is worthwhile to note in passing that despite the high degree of internationalization of many corporations in our sample, they still have one clear nationality (in one case two) in fundamental aspects of ownership, management, headquarter location, and cultural orientation. That is, the MNCs are in these aspects uninational with one exception, which is binational.

The companies were selected through panels of well-informed observers of Swedish, Japanese and US industry, respectively. Panel participants were asked to name those large manufacturing companies they considered to be in the forefront regarding technology management. Only three of these companies declined to participate.

In order to obtain a strategic perspective on technology management, the executive vice-president for technology (almost all Japanese corporations had such an executive) was interviewed in more than half of the cases. Interviewees for the rest of the companies ranged from the chairman of the board to the central R&D director. In addition to the interviews, a survey questionnaire was distributed among the companies, with a response rate of 81%. Part of this research is presented in this chapter.

The firms in this study account for a large part of the entire industrial R&D expenditures in their home countries: 38% for the 16 US companies, 21% for the 14 Japanese companies and 67.7% for the 12 Swedish companies.

It appears clear from the study that the concepts of multi-technology corporation and technological diversification have strong and increasing empirical relevance, e.g. as an important factor behind corporate growth (cf. [11, 13]). All of the companies in the study met ordinary requirements of being MTCs, that is, they were operating in at least three major technological areas. Moreover, the interviews indicated that 35 out of the 42 companies in the sample had diversified their technology base significantly during the last decade. More than half of the Japanese corporations had, as part of their corporate strategy, explicit plans for diversifying their products as well as their technologies, plans seldom found among US or Swedish corporations. Finally, a majority of the companies in the sample reported increasing R&D expenditures, increasing R&D intensity (that is, the R&D-to-sales ratio) and international activity in R&D.

9.1.4 Caveats

Many quantitative data in this chapter are based on perceptions by individuals in different companies and cultures, although with similar positions as technology executives. This imposes obvious limitations, which we cannot but urge the reader to keep in mind. One can argue extensively about the pros and cons of such an approach without any prospect of reaching a conclusive consensus. It is our belief that it is necessary but not sufficient to study perceptions, filtered in interviews and/or questionnaires, in research on strategic behavior and management, since the concept of strategic behavior has to include some degree of conscious *ex ante* determination by the agent (firm) of its future action. An assessment of strategic behavior based merely on past, observable behavior is therefore insufficient. Hopefully, indications emerge which together with case studies can be used for generating hypotheses which can be tested by more objective measurements (as has been the case regarding the phenomenon of technology diversification described in this chapter). Moreover, we believe that it is important to assess and analyze trends rather than optima or equilibria in evolutionary processes. Both by the choice of research questions, many of which are fairly novel, and by the choice of method, this chapter is exploratory and tentative. It can be fully justified only through further research.

9.2 TECHNOLOGY MANAGEMENT PERCEPTIONS IN JAPAN, SWEDEN AND THE USA

9.2.1 General Environmental Trends and Important Managerial Issues

As mentioned above, managerial perceptions are important when discussing and explaining corporate behavior. Perceptual differences reflect to some extent differences in the environments in which technology managers are working. The perceptions of what issues will affect the performance of the management teams, as perceived by technology executives in our questionnaire survey of MTCs, differ substantially between MTCs in the USA, Japan and Sweden, see Table 9.1.

Demand for higher quality emerges as the single most important issue. Keeping pace with technological progress also turns out to be of high priority. Issues associated with increased technological competition, such as shorter lead times and market lifetimes and more frequent introductions of new generations of existing products, turn out to be perceived as highly important by MTCs in the USA and Japan but not among managers in Swedish MTCs. Among Swedish MTCs, acquiring managerial talent, the influence of environmental pressure groups, and government intervention are perceived to be highly important for the performance of management teams. These findings are consistent with observations from executive interviews. Swedish technology executives seem to be less concerned with and worried by international competition, and to perceive government intervention

Internationalization and Diversification of MTCs 185

Table 9.1 Importance of managerial issues[a] and country differences as perceived by technology executives in MTCs in the USA, Japan and Sweden. N=26[b]

	Country diff.[c] 1988–1992			(Levels of significance)[d]		
	JPN N=11	SWE N=7	USA N=8	U/J	U/S	S/J
1 Keeping pace with new product technologies	2.82	2.33	2.75			.048
2 Fluctuating exchange rates	2.09	1.50	2.00			
3 Levels of exchange rates	2.18	1.50	1.75			
4 Acquiring managerial talent	2.09	2.33	1.62			
5 Low economic growth	1.27	1.67	1.88			
6 Government intervention	1.27	2.17	2.25	.013		.038
7 Coping with automation & computerization	2.64	1.67	1.88	.010		.019
8 Inflation	1.09	1.83	1.38			.032
9 Availability & cost of labor	2.09	2.00	2.00			
10 New competitors	2.00	1.50	2.38		.032	
11 Acquiring investment capital	1.60	0.67	1/50		.039	.029
12 Availability & cost of materials	2.27	1.50	1.50	.013		.026
13 Labor relations	1.54	1.17	1.88			
14 Trade barriers	2.18	1.17	2.00		.064	.008
15 Environmental pressure groups	1.18	2.33	1.38		.091	.005
16 Demand for shorter working week	1.64	1.33	0.50	.011	.039	
17 Escalating R&D spendings	2.27	1.17	2.12		.047	.006
18 Pressure for shorter innovation lead times	2.54	1.83	2.88		.016	.067
19 Shorter market lifetime of products	2.64	1.67	2.62		.041	.008
20 Pressure for more frequent introduction of new generation of products	2.45	2.00	2.50			
21 Pressure to acquire technology from outside company	2.09	1.17	2.12		.024	.023
22 Pressure for technological protectionism	1.82	0.83	1.88		.016	.035
23 Pressure for scientific protectionism	1.64	0.67	1.50			.033
24 Pressure to acquire technology from abroad	1.73	0.67	2.00		.008	.019
25 Increased complexity (fusion) of technology	2.70	1.67	2.50		.039	.006
26 Increased fusion between science and technologies	2.45	1.00	2.00		.022	.002
27 Demand for higher quality	2.82	3.00	2.88			
28 Level of interest rates	1.00	1.17	1.88	.048		

Scale: 0 = unimportant; 1 = of minor importance; 2 = important; 3 = of major importance
Notes:
[a] Principal issues concerning one major product area of each firm.
[b] Response rate = 62%.
[c] Country averages.
[d] Only significance levels below 10% are shown in the table.

and external non-industrial issues as much more important, compared with US and Japanese managers.

9.2.2 Discriminating Issues

These country differences are even more evident if we examine the complete set of perceptual data. A stepwise discriminant analysis ($R^2 = 0.74$) reveals that seven issues discriminate significantly among the three regions. Demand for a shorter working week (16) is seen as much less of a problem among US managers than among Swedish and especially Japanese managers. Pressure for shorter innovation lead times (18) is viewed as an issue of less importance among Swedish managers while US and Japanese managers believe this is an issue of major importance. Pressure to acquire technology from abroad (24), and the pressure for scientific protectionism (23), are perceived as very minor managerial issues among Swedish executives compared with US and Japanese executives. The levels of exchange rates (3) and the availability and cost of materials (12) are seen as much more of a challenge among Japanese managers, while Swedish technology executives experience problems with environmental pressure groups (15).

The Swedish technology executives in the sample perceive the environment as much less competitive and less of a potential performance problem than US and especially Japanese executives. Trade barriers related to goods, science and technology, and problems of international sourcing of technology, are seen as less of a problem among Swedish technology executives than among Japanese and US executives. The interviews support this pattern. One can only speculate about explanations for these differences. The Swedish companies are much more internationalized than their counterparts in the USA and Japan, and hence they have more international experience. Another factor behind this pattern is probably that Swedish companies are much more niche-oriented, and often dominate or are among the dominating firms within their niches, and hence have a better control over their immediate commercial environment. The problem in the longer run with this niche orientation or product specialization might be that the niches will be threatened by substitute products or technologies, which might lead to a lock-in effect.

9.2.3 Trends in Managerial Perceptions

A trend analysis of changes in managerial perceptions between two time periods, 1983–87 and 1988–92, reveals expectations for a more challenging environment globally as perceived by managers in the three countries. The following issues have gained significantly (on the 5% level or below) in importance between the two time periods among executives in all three countries: shorter lead times and market lifetimes, increased external acquisition (sourcing) of technology, technological protectionism, increased fusion of technologies, science with technology, and lastly increased demand for higher quality.

For US technology executives shorter innovation lead times, and external and international acquisition of technology are becoming more important.

Demand for higher quality has increased in importance among Swedish executives more than any other issue. Among Swedish technology executives the availability of labor and managerial talent is also perceived to be increasingly important.

Japanese executives experience increased importance of such competitive issues as shorter innovation lead times and market lifetimes, along with technology-related issues, such as external sourcing, technology fusion and fusion of science and technology.

9.3 INTERNATIONALIZATION OF TECHNOLOGY ACQUISITION

Several studies [7, 9, 19] have shown the increased tendency among large manufacturing firms towards external technology acquisition, a tendency confirmed in this study (see above).

The question now is whether not only externalization but also internationalization of technology acquisition in general takes place for large firms. There are no conclusive data yet on this score, only scattered indications that this is the case for most technology acquisition strategies. The following technology acquisition strategies have been distinguished in this study and ranked in order of decreasing degree of organizational integration (or equivalently in order of increasing degree of externalization). They are: *internal R&D, acquisition of technology based firms, cooperative R&D, technology purchasing and technology scanning.*

Regarding *internal* R&D, Table 9.2 shows significant country differences in the sample of Swedish, Japanese and US MTCs. The analysis of the share of R&D performed abroad in 1982 and in 1987 did not reveal (at 5% level) any significant changes in the percentage of internationalization of internal R&D. However, the interviews indicate that the share of R&D performed abroad will increase in many Japanese companies. Table 9.5 demonstrates that internationalization of R&D is becoming an important corporate strategy among MTCs in the USA, Japan and Sweden.

In a longer time perspective internationalization of R&D has certainly increased in many of the large Swedish companies studied. In Granstrand [3, Chapters 4–6] the historic pattern of corporate internationalization and internationalization of R&D was studied in seven large corporations in Sweden (Alfa-Laval, Astra, Boliden, Iggesund, Kema Nobel, SKF, Volvo) plus Philips. The corporations fell into two broad categories: product invention-

Table 9.2 Average percentage of R&D allocated centrally and in divisions, domestically and abroad 1987[a]

	JAP (N=14)	SWE (N=12)	USA (N=16)
Central/decentral. R&D (percentage split)	40/60[b,c]	8/92[b]	21/79[c]
Domestic/foreign R&D (percentage split)	97/3[d,e]	64/36[e,f]	86/14[d,f]

Notes:
[a] No significant global or regional changes in either centralization of R&D, or internationalization of R&D between 1982 and 1987 were found.
[b] Japan–Sweden difference significant at 1% level.
[c] Japan–USA " " 10% "
[d] Japan–USA " " 1% "
[e] Japan–Sweden " " 5% "
[f] Sweden–USA " " 8% "

based and raw material-based. Invention-based corporations had typically internationalized early, based on in-house R&D, while raw material-based corporations had typically diversified early but internationalized late. No corporation had been able to diversify and internationalize simultaneously due to scarcity of managerial competence, among other things. In most cases R&D and innovation had spurred corporate internationalization, and then the latter process subsequently spurred internationalization of R&D, which, however, became a feature first during and after World War II. Moreover, R&D was not initially internationalized as a result of a conscious corporate strategy but rather resulted from two primary factors[1]: local ambitions among managers and engineers in foreign subsidiaries, and acquisitions of foreign companies containing R&D [3, p. 113]. This is consistent with the findings in later studies of a larger sample by Håkanson, as presented in this volume.

Three stages of internationalization of the corporations were discernible, the third stage being characterized by multinational coordination of R&D, production and marketing through the use of matrix organizations, assignment of corporate responsibilities for R&D, production and supply to local centers and deliberate policies for enhancing local development for global markets [3, p. 121). The patterns of internationalization of firms in later stages had also become more varied and innovation-specific, which made models of ordered internationalization at firm level less valid (see also [27]). By the late 1970s, Alfa-Laval and especially SKF and Philips had reached

---

[1] Philips was a special case in that R&D personnel went to foreign subsidiaries to escape the Nazis. One may also note that some subsidiaries in Philips and SKF had started R&D in connection with contract work for the local national defence authorities.

this stage. These three corporations, founded in 1883, 1907 and 1891, respectively, had long experience of multinational operations going back at least to World War I. The first real multinational R&D organization was, however, Alfred Nobel's European network of R&D labs established in his dynamite trust in the 1880s. A detailed account of SKF's multinational coordination of R&D is given in [8][2].

All in all, in-house R&D among many large corporations in Sweden has become gradually internationalized since World War II, and R&D-based internationalization and product specialization have become their successful growth strategies, while product diversification has not.

Regarding *acquisition of technology-based firms*, Table 9.3 shows that the number and international dispersion of acquisitions in general among large, technology-based firms in Sweden have indeed increased. Lindholm [19] in a study of acquisition activities among large Swedish manufacturing firms in Sweden found that acquisitions of technology-based firms, made with the primary motive to source technology, have become more dispersed internationally even if the acquisition activities of Swedish firms are still guided towards Europe. (Cf. Håkanson in this volume who finds an increased activity among Swedish firms to establish R&D units in Europe, with acquisitions as a major route towards that end.)

Regarding technology *acquisition through external cooperative R&D*, three casual observations suggest that international acquisition of technology through external cooperation has become more common, at least in absolute terms. They are: an increase in the 1980s of technology-motivated strategic alliances in the industrial triad US–Europe–Japan, the creation of R&D subcontracting systems and the creation of international R&D consortia, especially in Europe. In Chapter 10 of this volume Mowery provides evidence for the increased internationalization of collaborative R&D especially among US firms. Additional support is presented by Chesnais [1].

Technology purchasing through licensing in has become internationally more dispersed since World War II (see [4]). International technology trade has not least been spurred by barriers to trade and foreign direct investment erected by many nations (China, USSR, India, Brazil, Japan, etc.). The massive flow of technology from the US to Japan during the 1950s to the 1980s through licensing is the most conspicuous case of international technology trade, by some called 'the biggest bargain ever' (see [16]). Considering this, it is an open question whether technology trade on an average has become more

---

[2] In passing one may note some peculiar features of SKF's corporate R&D laboratory set up in Holland in the early 1970s. It was set up with the primary purpose to function as a vehicle for coordinating R&D among SKF's subsidiaries, to which end a neutral location was chosen and a heavy joint investment made totally separate from local production and local market concerns and with a deliberate multinational staffing policy. SKF thereby became a company with highly internationalized R&D (the highest in Sweden in 1975) without any of the common motives behind the foreign R&D location.

Table 9.3 Acquisition activity among large Swedish manufacturing firms in 1977, 1982 and 1987[c]

	1977		1982		1987	
	Acq[a] (N = 51)	TAcq[b]	Acq (N = 59)	TAcq	Acq (N = 67)	TAcq
Total:	9	7	24	13	37	18
—in Sweden	3	2	9	6	8	5
—in other Nordic countries	0	0	7	2	5	3
—in other European countries	5	4	6	4	19	8
—in North America	1	1	1	1	1	0
—in Japan	0	0	1	0	1	0

Notes:
[a] Acq = total number of acquisitions made.
[b] TAcq = number of technology-based acquistions made for which technology sourcing was an important or very important reason.
[c] Source: [19].

internationalized in large firms in recent years. Purchasing of licenses is also the only external technology acquisition strategy that has not significantly increased in importance during the 1980s in any of our samples of Japanese, Swedish and US corporations. However, anecdotal evidence from technology traders, license brokers, etc. suggests that the volume of cross-border licensing is increasing, if not its share of all licensing agreements.

Technology scanning, finally, has for a long time been internationally oriented for both Swedish and Japanese companies, although several new countries have become additional targets for surveillance as they draw nearer the technology frontiers. In the 1980s foreign technology scanning gained increased attention in US companies (as did technology protection). Granstrand [5] reported from a study of 12 large US manufacturing firms that international technology scanning efforts had increased between 1980 and 1985, especially towards Japanese competitors. Also, various means (patents, secrecy, etc.) to protect proprietary S&T and intellectual property had increased in importance in the US [6].

9.4 DIVERSIFICATION

Granstrand, Sjölander and Alänge [13] argued that Swedish MTCs appear to be less product diversified and that their top management tends to put less emphasis on product diversification compared to their counterparts in US and Japanese MTCs. Granstrand and Sjölander [11] later showed that product *specialization*, internationalization and technology diversification formed a distinct success pattern among the 24 largest MTCs in Sweden between 1970 and 1985.

Table 9.4 shows how the technology executives in the country samples of large MTCs in the USA, Japan and Sweden assessed, on an average, the importance of different corporate strategies. The table shows that the importance of internationalization increased significantly between 1982 and 1987, and especially regarding internationalization of R&D. Internationalization is the strategy with the highest similarity among the countries. Significant country differences in the table confirm conventional wisdom which says that US companies emphasize profitability, while Japanese companies emphasize growth. Swedish companies put more emphasis on product specialization (the reverse of product diversification) rather than on product diversification as in the case of Japanese MTCs.

Almost all of the Japanese MTCs had diversified their technology bases significantly during the last decade, often in tandem with product diversification. However, even in cases of no product diversification, Japanese companies diversified their technology bases to a certain degree, due to increasing technological complexity in products and processes and advances in generic technologies (microelectronics, new materials, etc.) [13]. Also among Swedish companies there is a clear trend towards increasing technology diversification [20].

With respect to product diversification, especially the Swedish companies and to some extent the US companies specialize rather than diversify [13]. It is important to note here that in an analysis of product and technology diversification among the 21 largest Swedish manufacturing firms, no significant correlation between product diversification and technology diversi-

Table 9.4 Perceived importance of corporate development strategies—trends 1982–87

	General			Japan			Sweden			USA		
	82	87	p[a]	82	87	p[a]	82	87	p[a]	82	87	p[a]
Profitability	3.2	3.7	*0.03*[b]	3.0	3.5	0.17	2.9	3.4	0.18	3.5	4.0	0.17
Growth	3.1	3.2	0.62	3.4	3.6	0.40	2.6	2.7	0.69	3.2	3.1	0.83
Product diversification	2.8	2.5	0.21	2.9	2.9	0.98	2.6	1.7	0.18	2.8	2.5	0.32
Internat. of sales	3.0	3.3	0.12	3.3	3.1	0.41	2.9	3.1	0.52	2.9	3.6	*0.04*
Internat. of production	2.4	2.9	*0.05*	2.4	3.1	0.13	2.1	3.0	0.09	2.6	2.7	0.84
Internat. of R&D	1.6	2.6	*0.00*	1.7	2.7	*0.00*	1.4	2.9	*0.00*	1.5	2.4	0.07
Investments in R&D	3.1	3.2	0.75	3.2	3.5	0.38	2.6	2.7	0.79	3.2	3.1	0.62

Note:
[a] Level of significance.
[b] Numbers in italics mean that the correlate is significant on the 5% level or below.
Scale: 0 1 2 3 4; where 0 = of no importance and 4 = of major importance.

fication was found [20, p. 37], which indicates that technology diversification is not a result of product diversification.

Conclusive findings regarding the causes to and effects of these country-specific strategy differences are still lacking. Some hypotheses, however, are put forward by Jacobsson, Sjölander et al.

So far, Swedish companies have been studied in more detail by Oskarsson [20], who has shown strong and consistent evidence of increasing technology diversification at three levels of analysis in Swedish industry: the 4-digit ISIC level, the 21 largest R&D spenders, and three product areas—cellular phones, optical fiber systems and refrigerators. The result presented by Oskarsson is consistent with earlier research [11, 18, 21]. Furthermore, Oskarsson [20] investigated whether single categories of technologies cause the patterns of technology diversification. Electronics contributed significantly to these patterns at both industry and firm levels. However, all categories of generic technology diversification contributed at industry, firm and product level. This is a strong indication that technological opportunities are increasing on a wide front, as suggested by others [10]. Among the effects of technology diversification, Oskarsson supplied empirical evidence for the concept of multi-technology corporations, increased external technology sourcing, increased R&D costs, and a range of managerial challenges. The increased technology diversification was also found to be highly correlated with sales growth at firm level.

In the next section we will illustrate how technology, diversification and international competition are causally interlinked in one case: cellular systems and terminals.

9.5 INTERNATIONALIZATION AND DIVERSIFICATION: THE CASE OF CELLULAR SYSTEMS AND TERMINALS

9.5.1 Technological Progress, International Competition and Systems Development

In the spring of 1973 Ericsson's legendary technical director, Christian Jacobeus, initiated a technology scanning and feasibility study of mobile telephone systems. A group of people from Ericsson Telecommunications and Ericsson Radio Systems (at that time called SRA) was appointed to conduct the study. The recent progress in microelectronics, especially the development in LSI (Large Scale Integrated circuits), was considered to make it possible in the future to build mobile terminals and to develop systems with the needed capability.

The idea of making communication possible between mobile users, and between mobile users and the stationary worldwide telephone system, was certainly not new. In the late 1950s, Bell in the US had made experimental systems that consisted of a regular crossbar switch, a high-power radio base station (transmitter) to operate in a whole city, and bulky terminals mounted into cars and trucks. With this solution it was only possible to serve a couple of hundred subscribers in each area because of the limited number of available channels. This of course was a major commercial and practical problem among the very few and limited operating systems that existed in those early days of mobile communications.

In the late 1960s Bell invented the cellular system and developed the associated cellular technology to address the major operational problems experienced with the early mobile telephone technology. Up to 1982 Bell had invested more than $190 million without having a commercially sound operating system.

It was at Bell Labs that the transistor, which eventually led to the electronics industry, was invented in 1949 by William Shockley, who was awarded the Nobel Prize in 1956. It was Bell Labs that pioneered the development of microwave radio systems, and later, in 1962, its Telstar was the first active communications satellite in operation. Two important new transmission technologies and a generic technology (electronics) were added to the technology base of the telecommunications industry. It is a historical irony that those momentous new technologies of Bell Labs were eventually enlisted in the campaigns to undermine and finally dissolve the Bell System.

In 1980, the FCC (Federal Communication Commission) announced an important deregulatory decision. This decision drew a distinction between two kinds of services, basic and enhanced, which were differentiated by both technological (technologies nurtured at Bell Labs) and market characteristics. Basic services would remain under regulation while enhanced services, such as cellular systems, services and equipment, would be fully deregulated. The decision obviously led to quite a dramatic change, suddenly opening up the way to an avalanche of new suppliers of telecommunications products and services.

In a cellular radio system, the total geographical area to be covered is segmented into cells, each of which is served by a radio base station (RBS) of moderate power. Within each cell, radio transmission is achieved via a set of channels which can be reused in other non-contiguous cells. This frequency reuse is the key to efficient spectrum utilization and makes the cellular approach suitable for implementation of nationwide services. The radio network is tied to the public switched telephone network via special switching and control centers known as mobile switching centers, which in a modern system are based on stored program control technology. The more or less openly available cellular technology increases the number of possible subscribers by a factor of 100–500 in a large city, making commercial prospects considerably brighter and opening up the field for competition.

The prospects presented by cellular technology were certainly tempting for companies that believed they had the technological, financial and marketing capabilities required for capturing a substantial part of the market.

Ericsson approached the US cellular market through a series of marketing activities including participation in the Telecator exhibition in Hawaii, showing the NMT-450 system and the Saudi Arabian system during April and May 1982. At the end of May 1982 intensive lobbying was performed by Ericsson Radio Systems and the Ericsson Telecom people in Washington. More than 300 influential people listened to fairly dry, typical Swedish presentations. Ericsson was perceived as 'the serious alternative'.

A few days after Ericsson officially declared that they would go for the US non-wireline market, NEC, ITT/EF Johnson and Harris announced their intention to participate in the race for the non-wireline US market, although they did not have a complete reference system to show. In May Motorola began to take Ericsson's efforts seriously. Ericsson priced their switch slightly above Motorola and their RBSs slightly under, but most importantly Ericsson Radio Systems made their quotations in such a way that they could be directly compared to Motorola's.

When the non-wireline applications for the 30 largest cities were opened in June 1982, the Ericsson cellular system was specified in 45 out of 145 applications, while Motorola was specified in 90 applications. Harris and ITT were specified in only 5 applications. They were the true losers in this first round.

Motorola won the first two non-wireline *contracts* but their pricing and delivery dates were so tough that the operator in Buffalo who had specified Motorola in his application invited other suppliers to bid. The operator in Buffalo also went to Ericsson, and Ericsson Radio Systems won their first contract in May 1983 and succeeded in having that system in operation only 15 months later. This was the start of a successful venture for Ericsson even if there were some hard years during the mid-1980s, with weak economic results, severe quality problems in RBSs, scarce engineering resources and financial problems.

To secure the supply of RBSs and to expand their radio technology capability, Ericsson Radio Systems bought Magnetic, a major European manufacturer of RBSs, in 1984. The Swedish financial institution Investeringsbanken assisted in solving some of the major financial problems (they financed some of the major contracts) created by the tremendous growth and need for capital to put into R&D.

By 1990 Ericsson's US market share amounted to approximately 40% of the non-wireline systems market (20% of the total, non-wireline and wireline systems market). In Canada the same figures were 100% (50%)! Currently, approximately 45% of all subscribers in the world talk via Ericsson RBSs and switches, while the corresponding figures for Motorola and AT&T are approximately 25% and 20%, respectively.

Internationalization and Diversification of MTCs — 195

9.5.2 Systems Research and Development

When the quotations were made to potential US operators, Ericsson did not have a system. The development and adoption of the AXE switch and the radio base stations that were required were estimated by the R&D departments at Ericsson Telecom and Ericsson Radio Systems to need roughly 90 man years. Ericsson Telecom was to do 90% of the development work on the switch, meaning that Ericsson Radio Systems had to do roughly 40–50 man years of systems development. This had to be done in one and a half years. At that time they had 10 people engaged in the development of terminals and approximately 15 persons engaged in the development of the NMT 450 system and RBSs. They more or less had to triple their engineering manpower in a very short time period. This was done by extensive recruitment of engineers within the Ericsson group, and by attracting freshly graduated engineering students. Only five persons were left to do the development work on terminals (generation 1, see below) (mainly 'cosmetic' adjustment), while the remaining people were allocated to the development of switches and RBSs.

At the end of 1982, there was a dispute between the R&D people and the management of Ericsson Radio Systems. The management claimed that the mobile radio development needed six to seven people, while the R&D management said that the correct figure was 40 people. It is an after-fact that this in a way contributed to the success, because the management of Ericsson Radio Systems never really understood how complicated the development was, and therefore they committed themselves extensively to the US venture while underestimating the efforts required.

9.5.3 Development of Cellular Terminals[3]

Three generations of terminals have been developed at Ericsson; mobile analogue cellular phones (NMT 450), hand-portable analogue cellular phones (NMT 900) and digital hand-portable cellular phones (GSM). The external sourcing has increased for each generation. In 1989 Ericsson announced its US joint venture with General Electric, resulting in Ericsson-GE Mobile Communications. This joint venture was established to secure market channels and to support Ericsson's US market diversification. In 1990 Ericsson expanded its R&D activities in the cellular area to include a laboratory in the Research Triangle Park in North Carolina. This was done to contribute to Ericsson's technological capability, especially in the VLSI area.

In Table 9.5 the service characteristics, subtechnologies, diversification and sourcing pattern for the three generations of terminals at Ericsson are

---

[3] The results in this section are based on [20].

Table 9.5 The service characterisitics, subtechnologies, patent classes (IPC classes 4-digit level), technology diversification and sourcing for three generations of terminals at Ericsson

Product generation Service characteristics (product performance)	NMT 450	NMT 900	GSM
—Weight (in kg)	6	0.8	0.4
—Volume (in litres)	5	0.5	0.3
—Speech time (in hours)	Not relevant	0.4	3
—Speech quality	Low	Medium	High
—Speech security	Low	Low	High
—Interruption time length (seconds) (when changing cell area)			
—Inflation-adjusted price index	100	80	60
—Performance/Price ratio (Index/weight price)	17	100	150
Product subtechnologies			
—Analogue signal processing	Yes	Yes	No
—Analogue radio transmission	Yes	Yes	Yes
—Filter design	Yes	Yes	Yes
—Microprocessor application	Yes	Yes	Yes
—VLSI design	Yes	Yes	Yes
—Battery technology	No	Yes	Yes
—Liquid display technology	No	Yes	Yes
—Assembly design	No	Yes	Yes
—Surface mounting technology	No	Yes	Yes
—Polymer technology	No	Yes	Yes
—Digital radio circuitry	No	No	Yes
—Digital signal processing	No	No	Yes
—Digital to analogue converters (conversion of speech)	No	No	Yes
—Analogue to digital converters	No	No	Yes
—Software design	No	No	Yes
Technology diversification			
Number of technologies	5	10	14
Number of patent classes	17	25	29
Number of MSc categories with more than 15% of total engineer stock	1	3	4
Technology sourcing			
Number and share of the technologies supplied through external technology sourcing	0.6 12%	2.8 28%	4.0 29%
Number and share of the technologies supplied through international technology sourcing	0.3 6%	1.9 19%	2.1 15%
R&D costs			
R&D-costs index (Normalized man-year/generation)	100	200	500

Source: [20].

presented showing that the technology diversification between consecutive generations as well as external and international sourcing and R&D costs have increased.

9.6 INTERNATIONALIZATION, DIVERSIFICATION AND TECHNOLOGY—A TENTATIVE MODEL

As argued in the preceding sections, an important explanatory factor behind technological diversification is generic technological progress mainly within information technology, automation technology, new materials and biotechnology, along with increased technological competition. Granstrand and Sjölander [11] and Oskarsson [20] have shown empirical results that an important causal factor behind increased R&D spending is technological diversification. Technological diversification gives rise to an expanding opportunity set, as argued by Granstrand and Sjölander [10], which in turn means that the theoretically possible combinations of various technologies increase in number and are realized if technologies are combined or 'fused', to use Kodama's [17] terminology.

In Table 9.6 a correlation analysis is presented, based on managerial perceptions of statements associated with technological progress, technological diversification, competition (international, within industry, and from new entrants), and R&D (internationalization, international technology sourcing and R&D expenditures). The analysis attempts to relate technology diversification to technological progress, external sourcing, international industry competition, international sourcing and R&D expenditures.

The assertions about specific product areas which were presented to 26 technology executives of MTCs in the USA (8), Japan (11) and Sweden (7) are presented in the Appendix. The product areas incorporated major product areas of each MTC[4]. We can see in Table 9.6 that technology diversification is highly correlated with technological progress (information technology, new materials, automation technology), international competition (with new entrants, and companies from Japan, South East Asia and North America), external sourcing, international sourcing and R&D expenditures. Before interpreting the correlation pattern, let us return to the case of cellular systems and terminals to illustrate causality.

From Table 9.5 it is clear that the technological diversity, the degree of external as well as international technology sourcing, and the R&D costs have increased in subsequent generations of terminals. Despite the increased externalization and particularly internationalization of technology sourcing,

---

[4] The product areas were memory chips, optical fibers, cellular systems and terminals, PCs, TV sets, copiers, specialty steel, rolling bearings, robots, white goods and pharmaceuticals.

Table 9.6 Correlation analysis[a] of technological progress, technology diversification, internationalization, and competition (N=26)

Abbreviation[b]	1	2	3	4	5	6	7	8	9
(1) IntR&D	—								
(2) ExtR&D	0.35	—							
(3) IntTSourc	*0.39*	*0.71*	—						
(4) TDiv1	0.27	*0.68*	*0.45*	—					
(5) CompJ	0.20	*0.65*	*0.43*	*0.56*	—				
(6) CompSEA	0.17	*0.48*	*0.55*	*0.58*	0.02	—			
(7) CompNA	0.35	*0.43*	*0.53*	*0.50*	−0.26	0.08	—		
(8) CompE	0.26	*0.42*	0.16	0.03	0.20	0.24	0.34	—	
(9) CompNE	0.27	0.16	−0.03	*0.56*	−0.27	0.30	0.29	0.23	—
(10) IT	0.20	*0.56*	*0.39*	*0.53*	*0.42*	−0.03	0.04	0.32	*0.60*
(11) NewMat	0.25	*0.62*	*0.43*	*0.41*	0.29	0.25	*0.44*	0.33	0.09
(12) Bio	−0.16	0.02	0.11	−0.05	0.26	0.23	0.17	0.03	0.02
(13) AT	−0.11	*0.43*	*0.44*	*0.70*	*0.56*	0.03	0.11	0.30	*0.46*
(14) R&DI	0.16	−0.31	*0.41*	*0.49*	0.04	0.07	0.21	0.32	0.24

	10	11	12	13	14
(10) IT	—				
(11) NewMat	−0.13	—			
(12) Bio	0.14	0.29	—		
(13) AT	*0.44*	0.31	0.32	—	
(14) R&DE	0.36	0.18	0.02	0.35	—

Notes:
[a] Numbers in italics mean that the correlate is significant on the 5% level or below.
[b] IntR&D = international R&D, ExtR&D = external R&D, TDiv = technology diversification, CompJ, SEA, NA, E = Competition from Japanese, South East Asian, North America, European competitors, IT = information technologies, NewMat = new materials, Bio = bio technology, R&DI = R&D expenditures.

the increased number of technologies has caused a rapid growth in development costs. In order to build up a receiver capacity internal R&D was performed for technologies that were outsourced. Note that R&D costs increase progressively in relation to the increase in number of technologies. The data do not permit sensible curve fitting but it is quite conceivable that R&D costs would roughly rise with one component proportional to the number of technologies and one component proportional to the square of this number, reflecting coordination costs proportional to the number of binary interdependence relations in the set of technologies. The third generation (GSM) of cellular phones is about five times as expensive to develop (based on R&D man years) as the first generation (NMT 450). GSM-terminals are technically much more complicated than generations one and two, demanding close cooperative development work between skilled engineers from computer science, telecommunications signalling, man-made materials (plastics and displays), information theory and radio-technology disciplines.

Internationalization and Diversification of MTCs — 199

Furthermore, the development demanded intense cooperation with three Swedish universities of technology on digital signal processing.

The high development costs and the demand for multi-skilled people in R&D have also caused a clear tendency towards the emergence of strategic alliances for external technology sourcing and joint development. An example of this is the international strategic alliances in the development of GSM between Ericsson, Siemens, Matra and Texas Instruments (sourcing of high-performance integrated circuits), and between Nokia and AT&T.

What the case and the correlation pattern in Table 9.6 suggest is that when *generic technological progress* increases, the set of technologies possible to use in order to create the services that the customer is willing to pay for, also increases, whether the new technology is applicable to existing or new products or both. If the technological opportunities and the associated profit and growth prospects are to be realized, whether this is done through the development of new products (*product diversification*) or of existing products, *technology diversification* takes place. It is fair to say that in the case the generic progress within microelectronics and other information technologies spurred the diversity of technologies available. The future profit prospects of a fast-growing new product area attract new competitors as technology becomes more openly available, and this in turn increases competition, which contributes to technological progress. Ericsson had to source its technologies from outside Sweden, especially from the US. The cellular technology replaced land mobile radio technology and an expanding number of technologies were combined, through heavy R&D investments, to create the key user benefits of cheap and timely mobile communication. The correlation pattern in Table 9.6 shows an association between technological progress and technological diversification, international competition and sourcing.

Increasing technology diversification is indicated in this case by the expansion of the width of the engineering base, by the increased number of patent classes involved and by the resulting expanding set of subtechnologies in the product. When the first competitors reach the market with their products, competition increases, which in turn helps to expand the market because competition leads to higher performance and lower prices. As technologies become available on the international market, *international competition* through cost pressures, as well as pressure for shorter lead times in R&D and production, increases *external and international sourcing* of technologies and components (embodied technology). In the case presented here the technological diversification has increased the R&D costs faster than sales, resulting in increased absolute spendings as well as increased relative R&D spendings. Also the correlation pattern points at an association between R&D expenditures, on the one hand, and external sourcing, international sourcing and technology diversification, on the other. The reasons for increased external and international sourcing might be several. First, increased technology diversification changes the nature of the R&D activity into an

200

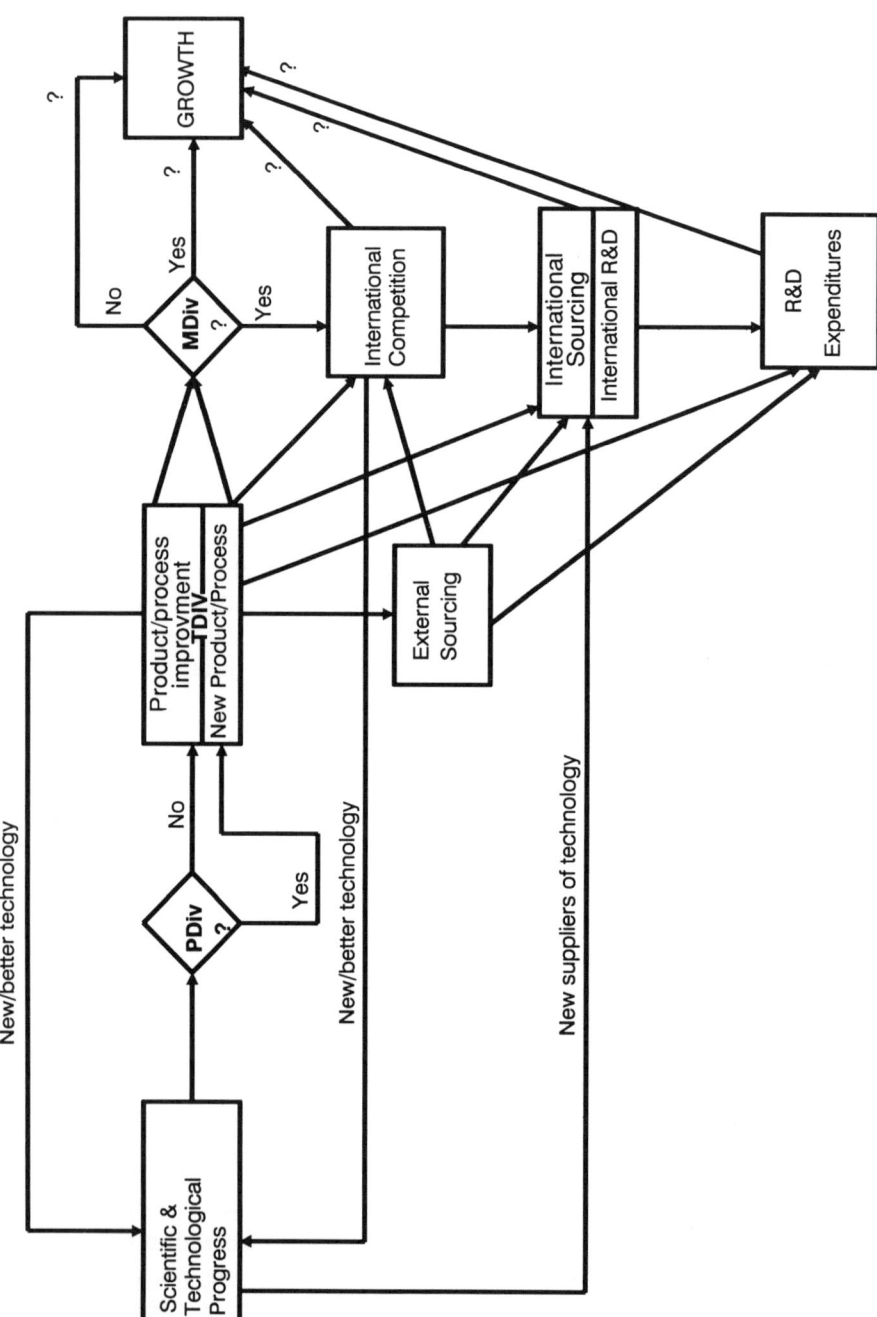

Figure 9.1 Internationalization, diversification and technology—a tentative model (PDiv = Product diversification, TDIV = Technology diversification and MDiv = Market diversification

increased number of technological areas, which naturally also demands an increased set of different technological competences. In the case study, technology sourcing was mainly externalized for new (to the firm) technologies that were so technologically complicated that they demanded a significant, and in some cases very large, amount of R&D if developed internally. Second, even if the firms had decided to develop all new technologies internally, finding sufficient R&D manpower would have been a critical problem (and maybe impossible to solve). Third, the externalization of technology sourcing appears to be facilitated by specialization among some technology-supplying firms. Many of the suppliers in this product area are increasingly specialized suppliers of sophisticated components and technologies for the product.

Companies have to make a choice—the choice of product specialization or diversification. Product specialization means an increased pressure for market expansion to create volumes large enough to pay for development (economies of scale). This can be accomplished through *market diversification* in the form of internationalization, or by increased penetration of current geographical markets. Product specialization also means that management attention is focused on fewer product areas, which, everything else equal, leads to higher effectiveness but also to higher vulnerability to rapid changes in market conditions and technology. On the other hand, the firm can choose to increase the range of product classes, that is increase its product diversity. In some cases this means that R&D investments can be distributed over more product classes to the extent that the technology involved is similar or the same and the firm has the ability to transfer technology internally. In the case described above, cellular terminals and systems are different products but have technological similarities and therefore product diversification means technological synergies, which if realized give economies of scope. Product diversification lowers the market risk exposure while product specialization does the opposite.

If we combine the process description of the development of cellular systems and terminals with the correlation pattern presented in Table 9.6, a tentative causal model can be hypothesized (see Figure 9.1). The model suggested in Figure 9.1 is obviously tentative and should be modified and tested in future research.

9.7 SUMMARY AND CONCLUSIONS

Between 1982 and 1987 internationalization has come to the forefront as the most important change in the strategies for corporate development. This appears to be caused by an increased pressure to expand the technology base, which in turn seems to be caused by the rapid development in such generic technology areas as microelectronics, information technology, automation technology, new materials and to some extent biotechnology, combined with the increased number of countries and companies pursuing high-quality R&D, contributing to an expanding set of technological opportunities. A strongly reinforcing factor is increasing R&D costs, which stimulate both increased supply of and demand for technology under certain conditions, e.g. that the technology supplier is not directly competing with the technology buyer and that the technology is offered at a price below the corresponding in-house R&D cost to the buyer. In general several factors may stimulate the supply of new technologies, such as: (1) high R&D costs for them at the supplier end in combination with lack of competition between supplier (which may be a product specialized firm or a university) and buyer (2) technological dependencies among companies, that is, companies need each other's technologies, and cross-licensing, patent pooling and other forms of swapping of technologies are possible and (3) the availability of even newer, substitute technologies at the supplying end. A competitive supply is moreover stimulated by (4) multiple sources of suppliers who believe that someone else is capable and willing to sell the new technology. ('If we don't sell, someone else will.')

There is, moreover, a clear tendency in our survey data towards increased importance of internationalization of R&D as a corporate strategy for technology sourcing. This is most probably also a function of an increasing global set of technological opportunities and the increased importance of external sourcing strategies. The probability of efficient sourcing of new technologies increases if a company has R&D activities geographically near their source. The interviews support the notion that the dominant reason among US and Japanese firms for internationalizing R&D is adaptation of products to local market needs. Another reason given among Japanese and Swedish companies is the need for local sourcing of technology, especially on the North American market [13, 16]. Output market conditions induce internationalization of R&D to follow the pattern of internationalization of sales and production. However, input (factor) market conditions—inducing R&D to internationalize according to the location of cost-effective supplies of inputs like materials, components, labor and especially knowledge—might increase in relative importance.

There is an almost prevalent phenomenon of increasing technology diversification at product, company and industry levels in Sweden (see [20]). Increasing technology diversification at company level is also positively correlated with growth of sales and growth of R&D (see [7]). Comparable quantitative data on technology diversification in Japanese and US companies are not yet available. However, the interviews in the sample of 14 Japanese companies indicated an increasing level of technology diversification in these companies as well. While Japanese companies diversify both products and technologies, Swedish and to some extent US companies are undergoing a phase of product specialization simultaneously with technology diversification. This suggests that the companies in the different countries make different trade-offs between economies of scope and economies of scale. While static as well as dynamic economies of scale (i.e. marginal cost reductions from production capacity and accumulated production respectively) favor product specialization, economies of scope favor product diversification, as well as technological diversification, everything else being equal (cf. [24, 25]). A scale-oriented management may then look upon internationalization and product diversification as a strategic choice, while a scope-oriented management regards them as complementary.

The question then arises whether Japanese companies are more suited and willing to reap economies of scope in R&D, production and marketing than Western companies. Certainly Japanese companies in the past have paid much attention to economies of scale, and learning-curve effects in particular. But there now seems to be a shift towards putting more emphasis on economies of scope. This may be partly due to technological advances, such as FMS and factory automation which may diminish both static and dynamic economies of scale, and partly due to managerial learning about how to manage R&D and multi-technology R&D in particular, taking advantage of specific features of the Japanese corporate society. Such features are longtime employment, low inter-firm engineer mobility, high intra-firm engineer mobility, rich communication behavior, substantial technology management efforts at the corporate level, agglomerate economies in the Tokyo and Osaka areas, government interventionism in pre-competitive stages, and weak territorial instincts among engineers.

We may *conclude* that in Japan, Sweden and the USA, there is a trend, to varying degrees, towards increasing technology diversification and increasing externalization and internationalization of technology acquisition. Technology diversification at the product level is likely to lead to increasing levels of R&D investments for each new product generation. This in turn puts pressure on companies to externalize and internationalize their technology sourcing, as well as to find different ways to exploit their technology. Product diversification could, moreover, offer ways to capitalize on in-house generic technological know-how. Increasing R&D costs, partly due to technology diversification, would then induce companies not only to pursue market diversification but also to expand the range of products with technical commonalities in order to recover the R&D investments. Altogether, this poses new challenges for managing R&D and technology efficiently. For example, coordination problems will increase further to the extent that R&D is internationalized and decentralized. The emergence of MTCs together with MPCs and MNCs (or the emergence of jointly diversified 'multi-firms') could be interpreted as one response to market failure. On the other hand, transactional cost considerations in the framework of Williamson [26] and Riordan and Williamson [22] lead to a mixed verdict between internal organization and market organization as the most efficient organizational form for conducting R&D and innovation (see [3], pp. 186–200). The emergence of quasi-integrated or intermediate organizational forms for technology acquisition and exploitation could then be interpreted as another response.

Thus, it is conceivable that linked to large technology-based firms, quasi-integrated internationalized systems for innovation would emerge as most conducive to innovation and technology-based growth, involving mixtures of internal R&D, subcontracted R&D, R&D consortia, alliances, acquisitions, internal venturing etc., all of which represent different degrees of management control and market mediation. However, the optimal (in some sense) degree of quasi-integration would in general be dynamically changing due to managerial learning and innovations (e.g. new contractual or organizational forms) and technological innovations (e.g. new information and communication technologies), which from time to time would tip the relative merits of internal versus market organization in one direction or the other by impacting transaction costs. If technological opportunities multiply rapidly in a combinatorial way, due to technology diversification, both markets and hierarchies would have difficulties in fully exploiting them, especially as long as managerial learning on how to handle this complexity is lagging behind.

In this context it must finally be noted that Williamson's framework of explanatory variables (uncertainty, bounded rationality etc.) behind transaction costs is not evolutionary in the sense that these variables may *explicitly* be affected by managerial learning or by new technologies that open up possibilities of improving management and/or market mechanisms. This does not prevent the original framework from being turned into such an evolutionary framework, but problems of operationalizing—and falsifying in a Popperian sense—Williamson's original variables then become compounded. However, rather than dismissing received transaction cost theory on these grounds, a search for a new framework of transaction cost determinants together with determinants of economies of scope (not only cost-based) and managerial learning might be rewarding in terms of better explanations and predictions of the continuing evolution of large, diversified corporations.

9.8 REFERENCES

[1] F. Chesnais (1988), Multinational enterprises and the international diffusion of technology, in Dosi et al. (eds), *Technical Change and Economic Theory*, Pinter, London, pp. 496–527.

[2] G. Dosi, C. Freeman, R. Nelson, G. Silverberg and L. Soete (eds) (1988), *Technical Change and Economic Theory*, Pinter, London.

[3] O. Granstrand (1979), *Technology, Management and Markets*, Department of Industrial Management and Economics, Chalmers University of Technology, Göteborg. Published in abridged and revised version by Pinter, London, 1982.

[4] O. Granstrand (1981), The Role of Technology Trade in Swedish Companies, Institute for International Business, Stockholm School of Economics, *Research Paper* No. 81/3 E, 1981.
[5] O. Granstrand (1986), Technology Intelligence in US Large Corporations. Working paper, Department of Industrial Management and Economics, Chalmers University of Technology, Göteborg, Sweden.
[6] O. Granstrand (1988), *Patents and Innovation*, CIM-Working Papers, WP 1988:04, Department of Industrial Management and Economics, Chalmers University of Technology, Göteborg.
[7] O. Granstrand, E. Bohlin, C. Oskarsson and N. Sjöberg (1992), *External Technology Acquisition in Large Multi-technology Corporations*, Department of Industrial Management and Economics, Chalmers University of Technology, Göteborg, *R&D Management*, **22**, No. 2.
[8] O. Granstrand and I. Fernlund (1978), Coordination of Multinational R&D, A Swedish Case Study, *R&D Management*, **9**, No. 1, 1–7.
[9] O. Granstrand, C. Oskarsson, N. Sjöberg and S. Sjölander, (1990), Business Strategies for New Technologies. Paper presented at the conference on Technology and Investment, in Stockholm, January, arranged by the Royal Swedish Academy of Engineering Sciences (IVA) in cooperation with OECD and The Swedish Ministry of Industry. Published in E. Deiaco et al. (eds), *Technology and Investment. Crucial Issues for the 1990s*. Pinter, London, pp. 64–92.
[10] O. Granstrand and S. Sjölander (1989), *Trender, problem och forskningsområden rörande teknikens industriella förverkligande—Speciellt Management of Technology*. CIM-Working Papers, WP 1989:04, Department of Industrial Management and Economics, Chalmers University of Technology, Göteborg (in Swedish).
[11] O. Granstrand and S. Sjölander (1990a), Managing Innovation in Multi-technology Corporations, *Research Policy*, **19**, No. 1, 35–60.
[12] O. Granstrand and S. Sjölander (1990b), The Acquisition of Technology and Small Firms by Large Firms, *Journal of Economic Behavior and Organization*, **13**, 367–386.
[13] O. Granstrand, S. Sjölander and S. Alänge (1989), Strategic Technology Issues in Japanese Manufacturing Industry. *Technology Analysis & Strategic Management*, **1**, No. 3, pp. 259–272.
[14] L. Håkanson (1989), Forskning och utbildning i utlandet. IVA-PM 1989:01. Royal Swedish Academy of Engineering Sciences (IVA), Stockholm (in Swedish).
[15] L. Håkanson and R. Nobel (1990), Determinants of Foreign R&D in Swedish Multinationals, Institute for International Business, Research Paper 9011, Research Policy.
[16] S. Jacobsson, S. Sjölander, O. Granstrand and S. Alänge (1989), *Strategic Technology Issues in US Manufacturing Industry—a critical analysis*, CIM-Working Papers, WP 1989:03, Department of Industrial Management and Economics, Chalmers University of Technology, Göteborg.
[17] F. Kodama (1986a), Japanese Innovation in Mechatronics Technology—Fumio Kodama studies technological fusion, *Science and Public Policy*, **13**, No. 1, 44–51.
[18] F. Kodama (1986b), Technological Diversification of Japanese Industry, *Science*, **233**, July 18, 291–296.
[19] Å. Lindholm (1990), Acquisition of Technology-Based Firms, A Study of Acquisition and Growth Patterns among Swedish Firms, Department of Industrial Management and Economics, Chalmers University of Technology, Göteborg.
[20] C. Oskarsson (1990), Technology Diversification—The Phenomenon, Its Causes and Effects, Department of Industrial Management and Economics, Chalmers University of Technology, Göteborg.

[21] K. Pavitt., M. Robson and J. Townsend (1989), Technological accumulation, diversification and organisation in UK Companies 1945–1983, *Management Science*, **35**, No. 1, January.
[22] M.H. Riordan and O.E. Williamson (1985), Asset Specificity and Economic Organization, *International Journal of Industrial Organization*, **3**, 365–378.
[23] P.P. Saviotti and J.S. Metcalfe (1984), A Theoretical Approach to the Construction of Technological Output Indicators, *Research Policy*, **13**, 141–151.
[24] D.J. Teece (1980), Economies of Scope and the Scope of the Enterprise, *Journal of Economic Behavior and Organization*, **1**, 223–247.
[25] D.J. Teece (1982), Towards an Economic Theory of the Multiproduct Firm, *Journal of Economic Behavior and Organization*, **3**, 39–63.
[26] O.E. Williamson (1975), *Markets and Hierarchies: Analysis and Antitrust Implications. A study in the economics of internal organization*, The Free Press, London.
[27] U. Zander (1991), Exploiting A Technological Edge—Voluntary and Involuntary Dissemination of Technology. PhD dissertation, Stockholm School of Economics, Stockholm.

APPENDIX: ASSERTIONS ABOUT TECHNOLOGICAL PROGRESS, TECHNOLOGY DIVERSIFICATION, INTERNATIONALIZATION, AND COMPETITION

Please respond to the following assertions about the future development of the product area by circling a number from −3 (= I strongly disagree) to +3 (= I strongly agree):

(1) IntR&D — We will perform our R&D abroad (in our own facilities) to a substantially greater extent.
(2) ExtR&D — We will source technology from outside our company to a substantially greater extent
(3) IntTSourc — We will acquire technology from abroad to a substantially greater extent.
(4) TDiv1 — Our technological capability (base of technology) will consist of a substantially larger number of technologies.
(5) CompJ — Competition from Japanese companies will increase.
(6) CompSEA — Competition from South East Asian companies (South Korea, Singapore, Taiwan, Malaysia, and Indonesia) will increase.
(7) ComNA — Competition from North American companies will increase.
(8) CompE — Competition from European companies will increase.
(9) CompNE — Competition from new entrants (new companies or companies entering our industry from outside) will increase.
(10) IT — The development within electronics, computer technology and information technology will affect our product area with substantially greater impact in the coming five years than in the past five years.
(11) NewMat — The development of new materials will affect our product area with substantially greater impact in the coming five years than in the past five years.
(12) Bio — The development within bio technology will affect our product area with substantially greater impact in the coming five years than in the past five years.

(13) AT   The development of various production technologies (robots, CAD/CAM, CIM etc.) will affect our product area with substantially greater impact in the coming five years than in the past five years.
(14) R&DE Our annual R&D (Research & Development) expenditures will increase substantially.

Chapter 10

International Collaborative Ventures and US Firms' Technology Strategies*

DAVID C. MOWERY

The number of international collaborative ventures involving US firms has grown significantly in recent years, and the activities included in these ventures also have shifted from a focus on manufacturing for a foreign market or extraction of natural resources to joint product development, marketing, or production for a global market. This chapter discusses the causes of growth and reorientation in international collaborative ventures involving US firms, noting the importance of change in the economic, technological, and political environment within which US firms now find themselves in global markets. The implications of international collaborative venture for US firms' technology strategies are also considered.

10.1 INTRODUCTION

During the past two decades the relative economic and technological strengths of the United States have declined. These strengths are now more evenly distributed among US and foreign firms. Combined with changes

---

*An earlier version of this chapter was presented as a paper at the Conference on Economic Growth and the Commercialization of New Technologies sponsored by the Technology and Economic Growth Program of the Center for Economic Policy Research at Stanford University, September 11–12 1989. I am grateful to conference participants for useful comments and suggestions. Preparation of the paper was aided by support from the Technology and Economic Growth program at the Center for Economic Policy Research at Stanford University, the Alfred P. Sloan Foundation, Berkeley, and the American Enterprise Institute.

Technology Management and International Business: Internationalization of R&D and Technology.
Edited by O. Granstrand, L. Håkanson and S. Sjölander. © 1992 John Wiley & Sons Ltd

in product and process technologies and in the industrial and trade policies of the United States and foreign governments, this new international economic environment has contributed to growth in the number and importance of collaborative ventures between US and foreign firms in the development and commercialization of new technologies.

International collaborative ventures are not completely novel undertakings for US firms, but those into which US firms have entered during the past decade differ in some significant ways from the joint ventures of the 1950s and 1960s. European and US participation in these ventures has grown simultaneously with expansion in domestic research collaboration in both the US and Western Europe. Examples of privately financed domestic cooperative research ventures in the US include the Microelectronics and Computer Technology Corporation (MCC) and others involving universities, such as the recently announced IBM–AT&T initiative in high-temperature superconductivity at MIT. In Europe, many regional cooperative research programs receive partial support from public funds, often under the sponsorship of the European Communities; current EC programs include ESPRIT, BRITE, RACE, and the older Airbus Industrie consortium, which is not directly sponsored by the EC and which focuses on product development and manufacture, rather than research. Interfirm cooperation in pre-commercial research, often supported in part with public funds, has been a prominent feature of Japanese science and technology policy during the 1960s and 1970s[1]. Much of this intra-Japan, intra-US and intra-EC collaboration is focused on activities that are somewhat further 'upstream' than those included in most of the international collaborative ventures that have developed in the last decade[2]. Most recent technology-centered international collaborative ventures appear to focus on product development, manufacture, and/or marketing.

International collaboration is in part a response to the same factors that have led to increased domestic collaboration in research among US firms. How have these factors contributed to growth in international collaborative ventures? How do they influence the types of collaboration between US and foreign firms that have developed in recent years? Has recent experience provided lessons for the organization and management of these ventures?

---

[1] See [42], [37] and [32] for discussions of U.S. and Japanese domestic research collaboration.
[2] In a discussion of collaborative ventures within the EC and those involving US and EC firms, Mytelka and Delapierre note that 'Intra-EC agreements are oriented far more towards knowledge production than are agreements between EC and American firms. Thus agreements embodying a knowledge component account for 41 per cent of total intra-EC agreements but only 20 per cent of the EC–USA agreements, and knowledge production agreements made up 21 per cent of intra-EC compared with 14 per cent of the EC–USA agreements. While intra-EC agreements stressed knowledge production, EC–USA agreements [emphasized] the need for greater market access. 37 per cent of the EC–USA arrangements involved the commercialization of products and a further 17 per cent the production of goods. The comparable figures for intra-EC agreements were 27 per cent and 9 per cent respectively...' ([35], p. 240).

International Collaborative Ventures and US Firms _____ 211

These issues are addressed in this chapter, which deals primarily with international collaborative ventures that focus on technology commercialization. This topic largely excludes detailed discussion of interfirm collaboration focused on pre-commercial or fundamental research, and omits interfirm collaboration that is confined to production or marketing.

10.2 INTERNATIONAL COLLABORATIVE VENTURES: DEFINITION AND GROWTH

Collaboration between US and foreign firms assumes numerous forms. Many collaborative ventures fit a narrow definition of a joint venture, including separate incorporation as an entity in which equity holdings are divided among the partners. Others, such as partnerships between 'risk-sharing' subcontractors and prime contractors or the purchase by one firm of an equity share in another, do not. In some industries, joint ventures and other forms of collaboration are extensions of subcontracting relationships that cover product development and manufacture; others focus on the marketing of products manufactured largely by one partner. This chapter uses the terms 'collaboration' and 'joint venture' interchangeably and therefore occasionally denotes as joint ventures organizational structures that do not fit the legal definition of this entity[3].

An international collaborative venture may be defined as *interfirm collaboration in product development, manufacture, or marketing that spans national boundaries, is not based on arm's-length market transactions and includes substantial and continual contributions by partners of capital, technology, or other assets.* In many cases, management responsibility is shared among the partner firms. This definition excludes other forms of international economic activity, such as export, direct foreign investment (which implies complete intra-firm control of production and product development activities), and the sale of technology through licensing. Nonetheless, in many cases, these various channels for the exploitation of firm-specific technological assets or capabilities are complements. Licensing in particular often is associated with collaboration.

One can distinguish among at least four types of technology-focused collaborative ventures[4]. The first involves collaboration among firms in research

---
[3] The formal structure of collaborative ventures, however, often has little to do with either their management or their success, as Gullander has noted: 'There are indications that the difference between a contractual and an equity relationship is highly exaggerated; "sophisticated" cooperators seem to downplay the importance of ownership control as compared to management control or control through other means.' ([17], p. 86). Porter and Fuller [40] make a similar point.
[4] Chesnais [11] provides a taxonomy of collaborative ventures on which I have drawn in defining these categories.

alone—these ventures and consortia, however, generally involve only domestic firms (defining 'domestic' to include firms from throughout the EC when discussing European domestic research collaboration). A second category of technology-focused collaboration that includes many international ventures is the exchange of 'proven' technologies within a single product line or across multiple products. These ventures have been particularly prominent in the global microelectronics industry, perhaps because of the long-established practice of cross-licensing within the industry, and are also widespread in robotics. A third category of international collaboration involves joint development of one or more products—this type of venture typifies international collaboration in commercial aircraft and engines and in some segments of the telecommunications equipment, microelectronics, and biotechnology industries. Finally, a number of international collaborative ventures in biotechnology, pharmaceuticals, steel, and automobiles involve collaboration across different functions, with one firm providing a new product or process for marketing, manufacture or application in a foreign market (or, in the case of biotechnology, marketing within the home-nation economy) by another. An important issue for future research concerns the extent of differences in stability or success among these forms of collaboration.

Joint ventures have long been common in extractive industries such as mining and petroleum production (see [44]) and account for a significant share of the foreign investment of US manufacturing firms since World War II[5]. Several features of recent collaborative ventures, however, differentiate them from earlier cases. The number of collaborations, both those involving only US firms and those between US and foreign enterprises, has grown. Joint ventures now also appear in a wider range of industries[6].

Since the early 1980s, other evidence suggests that growth in domestic and international joint venture activity has continued, and that recent inter-

---

[5] Hladik's analysis [22] of data from the Harvard Multinational Enterprise Project concluded that 39% of the number of foreign subsidiaries established by US manufacturing firms during 1951–75 were joint ventures.

[6] Harrigan [20] concluded that domestic joint ventures involving US firms had grown during the previous decade. In the 1960s, joint ventures were concentrated in the chemicals; primary metals; paper; and stone, clay, and glass industries, but now extend beyond these sectors. Hladik [22] found significant growth from 1975 to 1982 in the number of international joint ventures involving US firms, a growth trend that has almost certainly continued through the present. Hladik's conclusions disagree with those of Ghemawat, Porter and Rawlinson, who compiled a time series of 'international coalition announcements' (joint ventures, license agreements, supply agreements, and 'other long-term interfirm accords' [15], p. 346) covering 1970–1982 that displays no upward trend. The differences are likely to be more apparent than real; the Ghemawat et al. database includes a number of interfirm mechanisms of collaboration that are excluded by Hladik. The possibility thus exists that a shift in the mix of the different forms considered by Ghemawat et al. occurred during 1970–1982, as joint ventures and other collaborative agreement increased in importance relative to such alternatives as licensing.

national joint ventures continue to focus more heavily on technology-intensive activities (e.g. joint R&D, development or production) than the international joint ventures of the 1950s or 1960s. Hladik and Linden [23] found that between 1976 and 1987, the number of international R&D joint ventures entered into by US firms grew by more than 17% per year on average. By 1987, 47% of the ventures in their sample lay in the computer, electronics, semiconductor, or instrument industries. Analyzing domestic and international cooperative agreements by firms from the US, Europe, and Japan in biotechnology, information technologies, and new materials, Hagedoorn and Schakenraad [18] concluded that 'although technology cooperation between companies probably goes back many decades, it has experienced a major boost during the eighties' (p. 3)[7]. These authors found some slowdown in the rate of growth in new agreements in the late 1980s. Nonetheless, by the end of the 1980s, the number of new agreements being negotiated annually in the computer industry exceeded the annual number of agreements for the early 1980s.

The central activities of many recent collaborative ventures, including research, product development, and production for world markets, were absent from most of the ventures of the pre-1975 era, which focused primarily on production and marketing for the domestic market of the non-US partner firm. International collaborative ventures that deal with technology development and commercialization are based on an exchange relationship, but the relationship covers different commodities from those included in licensing or export transactions.

Some collaborations allow US and foreign firms to pool their technological capabilities in a single product without merging all of their activities into a single corporate entity. As was noted earlier, other forms of collaboration support the exchange of technologies across products. In most of these cases, the participants in the collaborative venture are competitors in one or more product markets. Still other collaborative ventures combine one firm's technological capabilities with the marketing or distribution assets of another for a single product. These ventures more frequently involve firms that are not direct competitors. Why are collaborative ventures preferred to exclusive reliance on alternative mechanisms, such as licensing, for these exchanges?

---

[7] Most of the data on which these analyses are based are drawn from the trade and financial press. The data suffer in varying degrees from at least two critical problems: (1) different research teams have utilized different definitions of international collaborative or technology-sharing agreements, and have employed different approaches to coding and categorizing the data; and (2) 'mortality' data rarely are available, making it impossible to determine whether new agreements are replacing or supplementing earlier ones. This last problem means that some portion of the measured growth in the number of international joint ventures during the past decade may be spurious.

10.3 COMPARING JOINT VENTURES AND OTHER CHANNELS FOR THE EXPLOITATION OF TECHNOLOGICAL AND OTHER FIRM-SPECIFIC ASSETS

In recent years, economists have emphasized the importance of firm-specific assets such as technological capabilities in explaining the growth of direct foreign investment and multinational firms[8]. Direct foreign investment, however, obviously is not the only mechanism for reaping the economic returns to these assets. Alternatives to direct foreign investment for the exploitation of firm-specific assets include licensing and collaborative ventures. As was noted above, these different channels for reaping the returns to an asset may be substitutes or complements, and firms frequently utilize multiple channels. The advantages and disadvantages of each of these methods depend on the characteristics of the asset in question. These factors also affect the activities undertaken within collaborative ventures, an issue that is discussed shortly.

Markets for the licensing of advanced technologies that exploit firm-specific knowledge often are very thin, with few buyers or sellers. The dearth of alternative outlets for sellers or alternative sources for buyers means that opportunistic behavior may hamper the operation of markets for technology licenses[9]. Other impediments to licensing technological assets include the tacit nature of much of the knowledge necessary to exploit the technology, the need to reveal a great deal about a technological advance in order to convince a prospective licensee of the value of the license ([2, 47]), and the problems of regulating licensor and licensee behavior in a dynamic and uncertain world.

Licensing is likely to be preferred to either direct foreign investment or collaborative ventures in technologies that are not complex, that have strong, well-enforced patents, that are relatively mature, and that do not rely on 'user-active' innovation (see [50]), with its requirements for strong links between marketing and product development. These features are not characteristic of the technology in the steel, automobile, commercial aircraft,

---

[8] Caves notes that '... as indicators of these [firm-specific] assets, economists have seized on the outlays for advertising and research and development (R&D) undertaken by firms classified to an industry. That the share of foreign subsidiary assets in the total assets of US corporations increases significantly with the importance of advertising and R&D outlays in the industry has been confirmed in many studies...' ([10], p. 9).

[9] In Williamson's terminology, a 'small numbers condition' characterizes such markets: 'The transactional dilemma that is posed is this: it is in the interest of each party to seek terms most favorable to him, which encourages opportunistic representations and haggling. The interests of the *system*, by contrast, are promoted if the parties can be joined in such a way as to avoid both the bargaining costs and the indirect costs (mainly maladaptation costs) which are generated in the process' [51, p. 26] (emphasis in original).

and telecommunications equipment industries, all of which have been active in international collaborative ventures. Licensing has been an important alternative to international collaboration in pharmaceuticals. Licensing and collaboration often complement one another in the microelectronics and robotics industries and in biotechnology.

Direct foreign investment provides the strongest basis for exploiting firm-specific technological capabilities with little danger of leakage or dilution of control. This intra-firm channel for exploitation of such capabilities involves high risks and costs, however, and may not facilitate rapid penetration of foreign markets. The cost penalties of establishing multiple production facilities argue against direct foreign investment in industries in which production technologies exhibit a large minimum efficient scale or strong plant-specific learning and cost-reduction effects; commercial aircraft (primarily airframes) and steel are examples of such industries. The high fixed costs and lengthy delays associated with establishing an offshore production, distribution, and marketing network may preclude direct foreign investment. Licensing, direct foreign investment, and export all force the innovating firm to bear all of the costs and risks of research and development, which have grown considerably in many US industries, especially those in the technology-intensive sector.

Uncertainties about economic and political conditions in many foreign markets may further reduce the attractiveness of direct foreign investment. Political barriers have impeded the establishment of wholly owned production facilities in high-technology industries in a number of nations; both the Japanese and the US governments, for example, have opposed some forms of direct foreign investment in their domestic semiconductor industries[10]. International joint ventures have substituted for, and in some cases have complemented, Japanese and US firms' direct foreign investment activities, especially in the integrated circuit, automobile, and pharmaceuticals industries.

What advantages does international collaboration have over licensing or direct foreign investment? Many of the contractual limitations and transactions costs of licensing for the exploitation of technological capabilities can be avoided within a collaborative venture. The problem of determining the value of partners' contributions can be reduced through collaboration. Partner firms make financial commitments to a collaborative venture that back their claims for the value of the assets they contribute; such financial commitments can substitute for the complete revelation of the value and

---

[10] Steinmueller [43] discusses the resistance of the Japanese government in the 1960s to the establishment by US firms of wholly owned manufacturing subsidiaries in microelectronics. US government opposition to the sale by Schlumberger of its Fairchild Semiconductor subsidiary to Fujitsu in 1984 is indicative of official US concern over Japanese investment in this industry. For further discussion of the influence of trade and technology policies on international joint ventures, see [30].

characteristics of the asset that may be necessary to complete a licensing agreement[11].

The noncodified, 'inseparable'[12] character of firm-specific assets that may preclude their exploitation through licensing need not prevent the pooling of such assets by several firms within a joint venture, or the effective sale of such assets by one firm to another within a joint venture. Joint ventures enable partner firms to 'unbundle' their portfolios of technological assets and transfer components of this portfolio, components that may be worthless in isolation, to a partner. The transfer of technology through a joint venture from a technologically advanced firm to a less advanced enterprise thus may enable the technologically 'senior' firm to reap some financial returns to portions of its portfolio of technological capabilities. The difficulties of unbundling the senior firm's technological portfolio for arm's-length transfer mean that these returns cannot be captured through licensing. Technology transfer may also be controlled or regulated more effectively within collaborative ventures than in licensing transactions[13]. Monitoring the behavior of the recipient of technology within a joint venture reduces the risk that the transferor will not benefit from any improvements in transferred technologies made by the recipient. These factors have been influential in collaborative ventures in steel (National Steel and Nippon Kokkan), automobiles (Toyota and General Motors), and commercial aircraft (General Electric and SNECMA; Boeing and the Japan Commercial Transport Development Corporation).

Collaborative ventures offer an alternative to the complete merger of firms as a means of pooling assets. Such ventures may cover only a limited range of products; partner firms often are competitors in other product areas. Collaboration can provide established firms a faster and less costly means than internal development to gain access to new technologies that are not

---

[11] Brodley summarizes the advantages of joint ventures, defined as separate corporate entities in which all partners hold equity shares, over mergers or market transactions as follows: 'By providing for shared profits and managerial control, joint ventures tend to protect the participants from opportunism and information imbalance. The problem of valuing the respective contributions of the participants is mitigated, because they can await an actual market judgement. The temptation to exploit a favored bargaining position by threatening to withhold infusions of capital or other contributions is reduced by the need for continuous cooperation if the joint venture is to be effective. Moreover, a firm supplying capital to the joint venture can closely monitor the use of its contributed capital and thereby reduce its risk of loss. Common ownership also provides a means of spreading the costs of producing valuable information that could otherwise be protected from appropriation only by difficult-to-enforce contractual undertakings. Finally, joint ventures can effect economies of scale in research not achievable through single-firm action. Because of these advantages, joint ventures are especially likely to provide an optimal enterprise form in undertakings involving high risks, technological innovations, or high information costs' ([6], pp. 1528–1529).

[12] 'Inseparability' refers to the fact that much of the firm's noncodified technological know-how may be embedded in the organization. Its transfer therefore will require the transfer of a large number of individuals. Separating and transferring a substantial portion of the parent firm's staff to another enterprise may be infeasible. (See [46] for further discussion.)

[13] Kogut [27] and Porter and Fuller [40] discuss the potential of joint ventures to reduce incentive conflicts and lower the possibilities for opportunistic behavior that often undercut licensing agreements.

easily licensed. This 'technology access' motive for collaboration between established and young firms has been particularly important in industries based on new technologies, such as biotechnology and robotics.

By comparison with direct foreign investment, licensing, or export, collaborative ventures also reduce the financial and political risks of innovation and foreign marketing. The products of a collaborative venture between a US and a foreign firm may well encounter fewer political impediments to access to the domestic market of the foreign firm than would direct exports from the US firm.

Nonetheless, the potential difficulties of collaborative ventures should not be minimized. Management of these undertakings has proven to be extremely difficult. Of the four categories of collaborative ventures described earlier, joint product development ventures appear to be the most complex and costly, occasionally resulting in either failure or significantly higher costs and longer development times than independent development. Even within a jointly financed technology development partnership, the value of partners' contributions may not be easily established, and this difficulty is compounded by uncertainty about technological and market outcomes in such a venture[14]. Ventures that are further removed from the market may face fewer difficulties in agreeing on the value of partner contributions, if the contributions (e.g. money or technical personnel) are truly homogeneous. Nonetheless, US domestic research consortia like MCC have had serious problems in extracting contributions of well-qualified personnel from participating firms (see [41]). These problems may be less serious in ventures that combine one firm's product development expertise with the production or marketing skills of another. Such ventures generally are closer to the market, which means that uncertainty about costs, prices, and volumes may be lower, and a value can be more easily assigned to the contributions. Misrepresentation of the value of these contributions nevertheless is possible in this situation as well, and the value of contributions is likely to change over time, forcing renegotiation or the demise of the venture.

Yet another important source of conflict in technology-focused collaborative ventures is technology transfer. Especially in ventures that involve firms with different technological capabilities, the senior firm wishes to minimize,

---

[14] Doz has pointed out the difficulties of measuring partner contributions in product development ventures that lie midway on a continuum between commercial production and fundamental research: 'When dealing with basic technology development—usually early in a partnership, much before a competitive stage—parity is maintained between the partners through balance in contribution; and the potential output is still so distant that precise valuation is not an issue. When dealing with well-developed technologies, a precise valuation of the outcome can be made, and the partners are almost at the stage of supply contracts, with precise products or system specifications, costs and prices, and some volume forecasts. Yet a "danger zone" often separates these stages in the evolution of a partnership, in the transition from precompetitive stages to competitive ones. During that transition one partner, but not the other, may shift from a valuation of the partnership based on contribution to one based on expected results, and show impatience with a divergence from the position of the other partner' ([12], p. 38).

and the junior firm to maximize, the amount of technology transfer. Even if conflicts among the participants over the amount of technology transfer can be resolved, actions to control transfer may threaten the viability of the venture.

This problem can be illustrated by the case of International Aero Engines, a consortium founded in 1982 to develop the V2500 jet engine that includes Pratt and Whitney, Rolls Royce, Fiat, Motoren-Turbinen Union, Ishikawajima-Harima Heavy Industries, Kawasaki Heavy Industries, and Mitsubishi Heavy Industries. Pratt and Whitney and Rolls Royce, the 'senior partners' within the consortium, attempted to minimize the transfer of engine technology within the consortium by designing the engine in modular form and assigning the development of different modules to different participants. Serious problems in the integration of the engine components, however, led to delays in the delivery of the V2500 engine and to a loss of orders ([3, 4, 9]).

The demands of product development partnerships also may clash with the other, independent activities or products of the participant firms. For example, independently manufactured products may become competitors with the jointly developed product. Such encroachment contributed to the 1977 demise of the collaboration of Pratt and Whitney and Rolls Royce that was intended to develop a high-bypass, high-thrust engine (the JT10D), and contributed to the collapse of a joint venture in airframes between Fokker and McDonnell Douglas in 1982.

Why have international collaborative ventures, which represent a hybrid of inter-firm and intrafirm modes for the exchange or sharing of technological and other assets, assumed greater importance recently for US firms, and why do these ventures now incorporate a wider range of activities? The basic answer is simple—changes in the technological and policy environment within which US firms operate have made the potential contributions of foreign firms to collaborative ventures much more attractive to US firms.

10.4 CAUSES OF INCREASED RELIANCE ON JOINT VENTURES

Changes in the technical capabilities of foreign firms and in the nature of product demand have increased US firms' demand for foreign partners in collaborative ventures. The enhanced technological capabilities of many foreign firms mean that their potential contributions to joint ventures with US firms now are more valuable. Foreign firms now are better able to absorb and exploit advanced technologies from US firms in industries in which there remains a substantial technology gap between US and foreign firms. In other industries, foreign firms either are more advanced or are the technological equals of US firms and therefore can contribute managerial or technological expertise to joint ventures with US firms. US firms in the automotive and steel industries and some US firms in the microelectronics industry now collaborate with foreign firms to gain access to superior foreign technologies.

The costs of the research and development necessary to bring a new product or process to market in many high-technology industries have risen considerably in the past 20–30 years—for example, commercial aircraft development costs have grown at an annual rate of nearly 20% for decades, despite advances in the application and productivity of the capital equipment used in the R&D process [31]. Similarly rapid growth in development and marketing costs has characterized the telecommunications equipment, computer, and microelectronics industries. Rising development costs place severe strains on the ability of firms to sustain ambitious R&D programs and increase the importance of penetration of foreign markets to ensure commercial success. Firms in some industries require markets substantially larger than those provided by a single Western European nation or even (in some cases) by the huge US domestic economy. Moreover, high development costs raise the risks of new product development, since they increase the fixed costs incurred before introduction of the product.

Another source of cost pressure on R&D programs is a form of technological convergence. Technologies that formerly were peripheral to the commercial and research activities of a firm now have become central to competitive advantage in a number of technology-intensive industries. The increased interdependence of telecommunications and computer technologies is perhaps the best known example of such convergence, but others include the growing importance of biotechnology within pharmaceuticals and food processing, or the greater salience of computer-based machine vision technologies within robotics equipment. Technological convergence means that firms must develop expertise quickly in a broader array of technologies and scientific disciplines, further straining R&D budgets and human resources. This factor has contributed to expansion in domestic inter-firm research collaboration and in research collaboration between industry and universities in the US and elsewhere[15]. Advanced communications and computer-assisted design and production technologies also facilitate the operation of international collaborative ventures.

A reduction in the duration of product cycles in many high-technology industries has increased the urgency of rapid penetration of global markets with new products. Rapid foreign market penetration is also more important because of the declines in the share of global demand accounted for by

---

[15] Granstrand and Sjölander [16] discuss this phenomenon of technological convergence in terms of the growth of 'multi-technology corporations', firms that either have to significantly broaden or overhaul their technology base. As these authors note, one means to acquire new capabilities is through a domestic or international joint venture.

the US market in many high-technology industries and the growing homogenization of demand characteristics across the industrial economies. The 'product cycle' model of direct foreign investment hypothesized that differences in local economic conditions gave rise to different firm-specific attributes that eventually were exploited in foreign markets through direct foreign investment. Increasingly, however, economic development and more rapid international technology transfer mean that the characteristics of domestic markets within the industrial world differ less, and the firm-specific assets and products that develop to serve these markets now are less 'country-bound' ([49, 13]). Simultaneous introduction of a product in multiple industrial economies now frequently is essential to commercial success. Such rapid penetration may require joint production or collaboration with a firm with an established marketing network.

The importance of technical standards for commercial success also makes rapid penetration of many markets with a new product particularly important in the microelectronics, computer, and telecommunications equipment industries. The establishment of a product as a *de facto* standard or dominant design may provide a profitable platform for the introduction of related products and subsequent generations of the dominant design[16]. In the global telecommunications equipment market, this dominant design motive is supplemented by the recognition that rapid penetration of many markets can contribute to a firm's influence within international standards negotiations. Collaborative ventures and technology exchange agreements now are pursued by some manufacturers of workstations and telecommunications equipment and services in order to achieve a dominant design position[17].

Still another factor underpinning the recent growth in both domestic and international collaboration involving US firms is the pivotal role of relatively

---

[16] The strong incentives to establish a product standard contributed to the practice of 'second-sourcing' in the semiconductor industry, which produced a complex network of technology exchange and cross-licensing agreements in the 1960s and 1970s. Farrell and Gallini [14] and Swann [45] analyze the monopolist's incentives to establish multiple sources for a new product. Second-sourcing was also directly encouraged by the Defense Department during the early years of the semiconductor industry. The incentives to establish one's product as a dominant design through encouraging duplication diminish somewhat the economic returns to strong intellectual property protection in these products. The shift in IBM's attitude toward imitators of the PS/2 from hostility and threats of patent infringement suits to liberalized licensing terms, illustrates the tradeoffs between intellectual property protection and strategies to establish a dominant design.

[17] See *Business Week*, which describes the licensing strategy of Sun Microsystems in workstations: 'Almost anyone can license Sun's basic software and Sparc, the superfast microprocessor that is the brain of its flagship workstation—a $9000-and-up desktop machine that packs the power of a minicomputer. If enough manufacturers build Sun clones, the software companies will have to take notice. In the end, everyone will prosper. And Sun's Sparc workstation—it makes six other models—will become a desktop standard...' ([8], p. 75; see also [24]).

International Collaborative Ventures and US Firms — 221

small startup firms in the commercialization of new technologies within the US (and therefore the world) economy. The successive waves of new product technologies that have swept through the post-war US economy, including semiconductors, computers, and biotechnology, have been commercialized in large part through the efforts of new firms[18]. Small firms appear to have been more important sources of new commercial technologies in the US than in Japan and Western Europe, where established firms have played a more significant role in new electronics or biotechnology products. In the microelectronics and computer industries, the important role of small firms resulted in part from US government procurement demand, which created a substantial market with comparatively low marketing and distribution barriers to entry. The benefits of the military market were enhanced further by the substantial possibilities for technological 'spillovers' from military to civilian applications.

The US military market no longer plays such a strategic role in the computer and semiconductor industries, and the possibilities for military-civilian technology spillovers appear to have declined in many areas of these technologies. (See [34] and [36] for further discussion of military-civilian technology spillovers.) Biotechnology firms also face markets that are heavily regulated in the US and other nations. As a result, the costs of new product introduction and the marketing-related entry barriers faced by startup firms in microelectronics, computers, and biotechnology now are higher. For this and other reasons, including the greater interest by foreign firms in the technological assets of US startup firms, collaborative ventures involving startup and established US and foreign firms have grown considerably in recent years. These ventures often focus on technology exchange (combined in many instances with the acquisition by an established firm of a substantial portion of the equity of the new firm) and/or marketing (including navigating domestic and foreign product regulations), rather than joint development of new products.

The growth of nontariff trade barriers[19] has also increased the incentives for US firms to seek foreign partners. Tariff barriers tend to favor direct

---

[18] This is not to deny the major role played by such large firms as IBM in computers and AT&T in microelectronics. In other instances, large firms have acquired smaller enterprises and applied their production or marketing expertise to expand markets for a new product technology. Nonetheless, it seems apparent that startup firms have been far more active in commercializing new technologies in the United States than in other industrial economies. Malerba's analysis of the evolution of the microelectronics industry in Western Europe and in the US [28] emphasizes the greater importance of startup firms in the US.

[19] See the estimate by Tyson [48] that 35% of US imports in 1983 were subject to nontariff restrictions, an increase from an estimated 20% in 1980. Olechowski estimated that 17–19% of the imports of developed nations (by value) were covered by nontariff barriers, and concluded that the use of nontariff barriers increased significantly during 1981–85 [38, p. 125].

foreign investment or joint production ventures as strategies for market penetration, since they affect only the relative prices of domestically produced and imported goods. Nontariff barriers, however, especially government procurement policies or technology transfer requirements, favor the use of collaborative ventures that incorporate product research, development, and marketing, as well as manufacture. Nontariff barriers in public procurement policies are significant in export markets for such goods as commercial aircraft and telecommunications equipment, where public ownership or control of major purchasers is common. Government procurement decisions can be influenced by the production (or development and production) of components for the purchased product by domestic firms in the purchaser nation. Foreign governments also frequently provide development funding and risk capital to domestic firms as part of industrial development policies. Combined with high product development costs, the availability of capital from public sources for foreign firms has enhanced their attractiveness as partners in product development ventures with US firms in microelectronics, commercial aircraft, telecommunications equipment, and robotics.

Just as foreign government trade and industrial policies have created incentives for US firms to collaborate with foreign firms in export markets, nontariff restrictions on foreign access to US markets have led to increased collaboration between US firms and foreign firms wishing to sell in the United States. In several protected US industries, a foreign production presence has been achieved through a joint venture. Examples include the Toyota–General Motors and Nippon Kokan–National Steel ventures. In the wake of the US–Japan Semiconductor Agreement and Fujitsu's failure to acquire Fairchild Semiconductor, joint ventures may become a more important means for Japanese semiconductor producers to establish a US production base (a similar argument is made in [5]).

Although the post-war growth of multinational firms and direct foreign investment raised the prospect in some assessments of 'global firms' to whom national boundaries would mean little or nothing, much of the current wave of international joint venture activity reflects the opposite phenomenon. National governments are able to influence not only production but, increasingly, the product development and technology transfer decisions of firms through the use of trade and other policies. Paradoxically, however, the pursuit by the United States and other industrialized nations of nationalistic or technologically mercantilistic policies of support for domestic industries has encouraged the development of consortia spanning national boundaries. The efforts of these governments to restrict international transfer of technology from publicly funded domestic research consortia that are closed to foreign firms often create strong incentives for the exchange of the fruits of these programs through international collaborative ventures [33]. This conjunction of nationalistic domestic research policies and expanding international technology transfer through collaborative ventures has been particularly pronounced in the global microelectronics industry[20].

10.5 INFLUENCES ON THE STRUCTURE OF INTERNATIONAL COLLABORATIVE VENTURES

One of the most important motives for international collaboration is access to markets, whether in the United States or in a foreign nation. The asset provided in exchange for market access frequently is technology. The form in which the technology is provided, which is determined by both the motives of the participants and the characteristics of the technology, plays a central role in structuring the collaboration. Thus, where the technology can be provided in essentially 'embodied' or codified form—i.e. in a finished product or a license—collaboration either is unimportant or focuses largely on marketing, as in the pharmaceuticals industry (outside of biotechnology).

In more mature high-technology industries, such as telecommunications equipment, microelectronics, and commercial aircraft, the high costs of new product development, demanding requirements for systems integration, and the nature of political barriers to market access all mean that many recent collaborative ventures have focused on product development. Because of the strong complementarities between Japanese firms' expertise in microelectronics process (CMOS) technology and US firms' expertise in product design for microprocessors, several of the US-Japanese collaborative ventures in microelectronics each span more than one product[21]. Demanding requirements for systems integration in robotics products that involve a widening array of technologies have also led to a number of domestic and international collaborative ventures among suppliers of robotics and factory automation equipment that deal primarily with product development. This recent wave of collaborative activity in robotics follows an earlier generation of user-supplier collaborations between robotics equipment suppliers and users in the automobile industry [26]. In commercial aircraft, telecommunications equipment, and segments of the microelectronics

---

[20] See [43, 11]. Chesnais has noted that an interesting complementary relationship may be developing between closed domestic research programs in the EC and the US, such as JESSI and Sematech, an international product development and technology exchange agreements in microelectronics: '...one finds a combination between *domestic* alliances in *pre-competitive* R&D (with all of the provisos attached to this notion), and a wide range of technology exchange and cross-licensing agreements among oligopolist rivals at the international level' ([11], p. 95; emphasis in original).

[21] Examples include agreements between Toshiba and Motorola and between Texas Instruments and Hitachi. See [43, 7].

industry, the desire of US firms for risk-sharing partners and for access to foreign capital or technology provides additional reasons to focus collaboration on research and product development.

As was noted earlier, much of the pervasive domestic and international collaboration in biotechnology focuses on the marketing and distribution by established firms of the technologies developed by new entrants. The market access motive for collaboration in this industry applies equally to domestic and international collaborative ventures, and domestic collaborative ventures therefore are more important relative to international ones[22]. Much of the domestic and international collaboration in this industry is intended to support entry into new domestic or foreign markets and does not always incorporate joint product development. The major international collaborations in the automotive and steel industries center on the exchange of foreign process technology, managerial expertise, and production systems for access to US markets. These joint ventures accordingly deal with production rather than with product development.

10.6 MANAGING INTERNATIONAL COLLABORATIVE VENTURES

Large-scale databases on international collaborative ventures' duration, success and failure do not yet exist to allow empirical testing of detailed hypotheses about the factors that contribute to the success and failure of these ventures. Nonetheless, the available empirical studies and a growing body of case study and anecdotal evidence suggest a number of factors that should be addressed in the organization and management of international collaborative ventures that are technology-focused.

Management of these undertakings should be premised first of all on a recognition of their dynamic character. Firms' motives for collaboration often center on knowledge or technology acquisition—once these processes are sufficiently advanced or completed, partners may no longer wish to remain in a joint venture. These motives themselves often change in response to changes in the environment or within the participating firms.

---

[22] Hagedoorn and Schakenraad concluded from their analysis of publicly announced collaborative agreements that international collaborative ventures were more important relative to domestic (including intra-EC) ventures in information technology than was true of biotechnology: 'Over 12% of all biotechnology agreements are intra-European agreements and almost 17% are between European and US companies. In information technology the number of European-US agreements comes very close to the number of intra-US agreements. Intra-European agreements reach 17% of all such agreements' ([18], p. 15).

The value of the assets contributed by participating firms to the collaborative venture may also shift in response to changes in markets or technology. Virtually all of these statements apply as well to a wholly-owned corporate subsidiary or foreign investment, but the presence of other corporate actors in a joint venture increases the potential for rapid and dramatic change.

A second important precondition for management of joint ventures is a realistic assessment of their costs and benefits—in recent years, the latter may have received more attention than the former. The costs, risks, and benefits of alternatives should be compared with those of a collaborative strategy venture, and comparisons should be made among alternate structures for a collaborative venture. As was noted earlier, technology-focused collaborative ventures that center on joint development of a product appear to be more difficult to organize and manage than other forms of international technology-centered collaboration. Alternative structures for interfirm collaboration, including technology exchange agreements, collaboration that is restricted to research, and marketing or production agreements, deserve careful scrutiny and comparison.

Since technology transfer is at the center of many joint ventures, the management by partner firms of both technological development and technology transfer is critical to the success or failure of international collaboration. The interests of technological leaders, reluctant to allow the transfer of key technological capabilities, differ from those of technological followers, whose participation often hinges on the amount of technology transfer. The success of a project may be undercut by the attempts of 'junior' partners, motivated by their desire to maximize technology transfer and learning, to participate in all aspects of the project, rather than specializing in a particular area or activity. Alternatively, as was noted earlier, efforts by the senior partners to restrict technology transfer create two risks—alienation of the junior partner firm(s) and/or failure of the project to reach its technical goals.

Entry into a collaborative venture therefore must be predicated on an expectation that technology will be transferred to one's partners. But how much and which technologies? Participation in a collaborative venture requires that the firm assess its own technological capabilities and distinguish between those that are critical to its competitive performance and should not be transferred and those components of the corporate technology portfolio that can be transferred profitably with less risk to the competitive performance of the firm. Two other assessments are also necessary. The firm contemplating entry into a collaborative venture must also examine the non-technological, complementary assets that are needed to realize the commercial returns to its technological capabilities. As Teece [47] and others have shown, access to these nontechnological assets can determine innovative success or failure. Finally, the firm contemplating a collaborative venture must examine the technological and nontechnological capabilities of its prospective partners in a collaborative venture before the inception of a venture, because of the central importance of choice of partner(s). In all of these assessments, one must recognize that the quality and abundance of the key technological and nontechnological assets may well change over time.

Technology transfer must be managed carefully over the life of a collaborative venture[23]. Participant firms should create internal mechanisms for absorbing technology transferred from other partners. These may involve the regular rotation of personnel through a collaborative project. The importance of personnel flows reflects the tendency for technology transfer to operate more effectively through the movement of people, rather than through a flow of reports and paper. It is not enough, however, to simply capture knowledge (both codified and tacit) or skills from other firms in the head of an engineer or scientist—that individual must be given opportunities to communicate such knowledge to others within the parent firm. Parallel research or engineering activities within the parent firm are used by many firms in university–industry research collaborations to provide an informed audience for the technological and other knowledge gleaned from a collaborative venture. They may be equally useful in a collaborative venture that focuses on commercial technology development. Monitoring the flow of technology transfer also may be difficult in large firms engaged in a collaborative venture that spans several business units. In these circumstances, a central office or point of contact can monitor the requests for technological or other information and data from partner firms to all divisions or subsidiaries of the corporation, enabling senior managers to track the 'balance of trade' within the venture more effectively.

Despite the complexities that it creates for management, technology transfer often acts as a source of cohesion within product development ventures, especially those involving a dominant and a subordinate firm. These ventures often are more durable and successful than those among technological equals [25]. Evidence from the commercial aircraft industry, in which international collaboration in development projects has been widespread for more than a decade, strongly supports this hypothesis. Product development ventures of technological equals, such as those between Rolls Royce and Pratt and Whitney in the JT10D jet engine project, Fokker and McDonnell Douglas in the MDF100 commercial aircraft project, and Saab and Fairchild in the SF340 commuter aircraft project, repeatedly have failed to bring a product

---

[23] Several of the arguments in this paragraph are also made in [19].

to market or have been unable to achieve commercial success with a product after its introduction. Product development ventures between technologically dominant and subordinate firms, however, such as the CFM International venture between General Electric and SNECMA of France and the collaborative ventures involving Boeing and the Japan Commercial Transport Development Corporation appear to be more manageable.

Technology transfer has also acted as an adhesive rather than a solvent in collaborative ventures that span several products in the microelectronics industry. Monitoring of technology transfer and the clear establishment of benchmarks for reciprocity within a venture may be easier when one firm's proprietary technology in one product is being traded for a partner's expertise in another product. This form of reciprocity appears to have aided the Motorola–Toshiba venture, in which Toshiba's CMOS process expertise is being exchanged for Motorola's microprocessor design capabilities in an enterprise that produces both microprocessors and DRAMs in a Japanese facility ([19, 1]).

The dynamic character of collaborative ventures is reinforced by a tendency for the assets contributed by each partner firm to gradually lose their value to the other participants. As technology is transferred through a collaborative venture, learning by the other participants will reduce the value of the technological capabilities that originally were unique to one or another participant. Depreciation is likely to be even more rapid in ventures in which one firm contributes its marketing knowledge and network or other 'country-specific' expertise—as the other participants improve their knowledge of the markets in which this partner has specialized, they may well choose to continue without it[24]. Depreciation in the value of US firms' contributions has played a role in the breakup of a number of collaborative ventures formed between Japanese and US producers of auto parts. As the Japanese partners in these enterprises gain knowledge about local markets and (particularly when selling to Japanese transplant operations in the US) local production conditions, they frequently withdraw from the joint venture to continue independently, as Phillips [39] has noted. Although technology-based assets are likely to depreciate more slowly, especially if technology transfer is closely managed, Hamel, Doz and Prahalad [19] suggest that process technologies are less transparent to other participant firms and therefore may depreciate more slowly than product technologies, which venture partners may learn more easily[25].

---

[24] Porter and Fuller have observed that collaborative ventures centered on marketing 'may be particularly unstable, however, because they frequently are formed because of the access motive on one or both sides. For example, one partner needs market access while the other needs access to product. As the foreign partner's market knowledge increases, there is less and less need for a local partner' ([40], p. 334).

Depreciation in the value of assets within a joint venture is no less inevitable than depreciation of physical capital assets within a factory. In both cases, participants must take steps to reduce erosion in the value of their contribution. Intra-firm technology development must underpin the technologies contributed to the joint venture; where a firm is providing a 'static' asset like market access, the collaborative venture may function most effectively as a means for exit from the industry or as a channel for learning process and product technologies.

The organizational structure of international collaborative ventures, especially those involving joint research and product development, raises additional challenges. There is no optimal management structure for a collaborative venture—its design will depend, among other things, on the character and magnitude of the contributions of the participants. In collaborations of technological equals, an autonomous management structure in charge of a wide range of design, marketing, production, and product support may be preferable. Such a management structure often is costly, since it duplicates some or all of the management structure of the member firms. Nevertheless, the experience of recent collaborative ventures clearly indicates the importance of strong links between the product development and design team and the organization responsible for marketing and product support. The organization managing the collaborative venture, be it the single dominant firm or an independent hybrid of the parent firms, must retain control of a number of downstream activities. On the other hand, in collaborations involving a senior and a junior firm, financial and organizational structure appears to be less important, so long as the technologically more advanced firm retains overall control of technology and management decisions.

Finally, the case of the Anglo–French Concorde partnership, where total project costs rose from $450 million in 1962 to $4 billion by 1978, illustrates the need for building cost controls into the structure of a collaborative venture. This issue is important because of the tendency for participants to be less concerned about minimizing shared costs. In the Airbus and other joint ventures with less disastrous financial consequences than Concorde, fixed-price contracts between a central management organization and the

---

[25] The type of skill a company contributes is an important factor in how easily its partner can internalize the skills. The potential for transfer is greatest when a partner's contribution is easily transported (in engineering drawings, on computer tapes, or in the heads of a few technical experts); easily interpreted (it can be reduced to commonly understood equations or symbols); and easily absorbed (the skill or competence is independent of any particular cultural context).

'Western companies face an inherent disadvantage because their skills are generally more vulnerable to transfer. The magnet that attracts so many companies to alliances with Asian competitors is their manufacturing excellence—a competence that is less transferable than most' [19, p. 136]. Of course, the reverse is also true—a central technological asset contributed by Boeing to its collaborative ventures with Japanese firms is the US firm's expertise in production technology and in the management of fluctuations in production volume for commercial airframes [29].

partner firms have preserved incentives for partners to minimize costs. Profit-sharing, rather than cost-sharing, is crucial.

This discussion provides additional reasons to be skeptical about the prospects for collaborative ventures in product development. Within this class of collaborative ventures, those involving partnerships of technological equals appear to be the most difficult. In any consideration of collaborative ventures, then, the product development collaboration of equals should receive the most severe scrutiny and critical analysis. Another issue that spans many of the factors discussed above concerns the relative stability of partnerships of firms with similar assets and the durability of ventures involving firms with complementary capabilities or assets. Despite assertions to the contrary in much of the literature, this analysis suggests that within technology-centered collaborative ventures, complementary capabilities are a greater source of strength and stability than are strong similarities in the technological and other assets contributed by partner firms to a collaborative venture[26]. As in other endeavors, opposites may attract in international collaborative ventures, and the resulting relationship may be more durable than one based on similar endowments.

10.7 CONCLUSION

International collaborative ventures are a relatively new channel for US firms' technology development and commercialization efforts. Their very novelty makes it difficult to render a definitive verdict on their implications for national competitiveness, but current evidence does not support the view that they will do great damage. Any summary evaluation of collaborative ventures must recognize the great differences among industries in the focus and structure of international collaboration, as well as the differences in the international flows of technology that occur within these undertakings. In most cases, international collaborative ventures are a result, rather than a cause, of intensified international competition with the products or technologies of US firms. They reflect changes in the international competitive environment (e.g. the reduction of US technological and economic hegemony as a result of economic reconstruction in Western Europe and Japan, combined with economic development in East Asian economies) that long were the objectives of US foreign and economic policies.

Equally difficult to summarize from the limited experience with these ventures are the optimal approaches to organizing and managing these ventures. Careful consideration of collaborative ventures must recognize their

---

[26] Both Porter and Fuller [40] and Hennart [21] argue that partnerships based on similar assets are more durable, although Hennart applies his observation to a broader class of collaborative ventures than those centered on technology.

complexity and evaluate alternatives to collaboration. Many of the policies suggested earlier, such as the assessment of the technological and non-technological assets that are essential to a firm's competitive performance and those of one's competitors, are also necessary for the management of technology development and commercialization within the firm. Recognition of the dynamic nature of collaborative ventures and of the need to invest internally in the strengthening of the assets contributed to the joint venture and in the transfer and exploitation of learning within the joint venture, also is important. International collaboration may be a less costly and faster means of commercializing new technologies, but it is by no means costless.

Substantial additional research is necessary in order to formulate a richer set of prescriptions for public policy makers and private managers. In particular, better data on the structure and performance of collaborative ventures are essential. These data will not be easily collected from the trade and financial press, the source of many of the datasets that have been used thus far in empirical work on international collaborative ventures. It is also important to broaden the data to include firms that are not among the ranks of the global multinationals—international collaborative ventures increasingly will affect US firms that historically have not been involved in international operations. Finally, more research is needed on the microeconomic causes of increased domestic and international collaboration in research and technology development. We know relatively little, for example, about the causes (or even the rate) of escalation in the costs of technology development and commercialization, although this escalation surely contributes to growing collaboration in the international and domestic spheres. The phenomenon of technological convergence is also widely remarked but rather less widely measured, analyzed, or understood. Better understanding of the causes and implications of international collaborative ventures will require analysis of these and other conditioning factors.

10.8 REFERENCES

[1] L. Armstrong (1988), A Chipmaking Venture the Gods Smiled On, *Business Week*, 4 July, 109.
[2] K. J. Arrow (1962), Economic Welfare and the Allocation of Resources for Invention, in Universities-National Bureau Committee for Economic Research, *The Rate and Direction of Inventive Activity*, Princeton University Press, Princeton, N.J., 1962.
[3] *Aviation Week and Space Technology* (1987), U.S., Europeans Clash Over Airbus Subsidies, 9 February, 18–20.
[4] *Aviation Week and Space Technology* (1987), Pratt and Whitney Expands Role in V2500 Compressor Work, 16 March, 32–33.
[5] M. Borrus (1988), *Competing for Control*, Ballinger Publishers, Cambridge, 1988.

International Collaborative Ventures and US Firms — 231

[6] J. Brodley (1982), Joint Ventures and Antitrust Policy, *Harvard Law Review* **95**, 1523–1590.
[7] *Business Week* (1989a), Is the U.S. Selling Its High-Tech Soul to Japan?, 26 June, 117–118.
[8] *Business Week* (1989b), Clonemakers Don't Scare Sun—It's Sending Them Engraved Invitations, 24 July, 75.
[9] W.M. Carley (1988), Cancelled Jet Order is a Setback for United Technologies Unit, *Wall Street Journal*, 2 September, p. 6.
[10] R.E Caves (1982), *Multinational Enterprise and Economic Analysis*, Cambridge University Press, Cambridge.
[11] F. Chesnais (1988), Technical Co-Operation Agreements Between Firms, *STI Review* **4** (1988) 51–119.
[12] Y.L. Doz (1988), Technology Partnerships Between Larger and Smaller Firms: Some Critical Issues, *International Studies of Management & Organization* **17**, 31–57.
[13] J.H. Dunning (1988), *Multinationals, Technology, and Competitiveness*, Unwin Hyman, London.
[14] J. Farrell and N. Gallini (1988), Second-Sourcing as a Commitment: Monopoly Incentives to Attract Competition, *Quarterly Journal of Economics* **103**, 673–694.
[15] P. Ghemawat, M. E. Porter and R. A. Rawlinson (1986), Patterns of International Coalition Activity, in M. E. Porter (ed.), *Competition in Global Industries*, Harvard Business School Press, Boston, Mass.
[16] O. Granstrand and S. Sjölander (1990), Managing Innovation in Multi-technology Corporations, *Research Policy* **19**, 35–60.
[17] S. Gullander (1986), Joint Ventures in Europe: Determinants of Entry. Working paper, Columbia University Graduate School of Business.
[18] J. Hagedoorn and J. Schakenraad (1988), Strategic Partnering and Technological Co-operation, presented at the EARIE conference, Rotterdam, 31 August–2 September.
[19] G. Hamel, Y. Doz and C.K. Prahalad (1989), Collaborate with Your Competitors—and Win, *Harvard Business Review*, January–February, 133–139.
[20] K.R. Harrigan (1984), Joint Ventures and Competitive Strategy. Working paper, Columbia University Graduate School of Business.
[21] J-F. Hennart (1988), A Transactions Cost Theory of Joint Ventures, *Strategic Management Journal*, **9**, 361–374.
[22] K. Hladik (1985), *International Joint Ventures*, D.C. Heath, Lexington, Mass.
[23] K. Hladik and L.H. Linden (1989), Is an International Joint Venture in R&D for You?, *Research & Technology Management* **32**, 11–13.
[24] J. Khazam and D. Mowery (1991), RISC: Rewriting the Dominant Order in the Microprocessor Industry, unpublished MS, University of California, Berkeley.
[25] J.P Killing (1983), *Strategies for Joint Venture Succes*, Praeger, New York.
[26] S. Klepper (1988), Collaborations in Robotics, in D.C. Mowery (ed.), *International Collaborative Ventures in U.S. Manufacturing*, Ballinger Publishers, Cambridge.
[27] B. Kogut (1988), Joint Ventures: Theoretical and Empirical Perspectives, *Strategic Management Journal* **9**, 319–332.
[28] F. Malerba (1985), *The Semiconductor Business: The Economics of Rapid Growth and Decline*, University of Wisconsin Press, Madison, WI.
[29] D.C. Mowery (1987), *Alliance Politics and Economics: Multinational Joint Ventures in Commercial Aircraft*, Ballinger Publishers, Cambridge, Mass.
[30] D.C. Mowery (1991), Public Policy Influences on the Formation of International Joint Ventures, *International Trade Journal*, **6**(1), 29–62.
[31] D.C. Mowery and N. Rosenberg (1982), Government Policy and Innovation in the Commercial Aircraft Industry, 1925–75 in R. R. Nelson (ed.), *Government and Technical Change: A Cross-Industry Analysis*, Pergamon Press, New York.

[32] D.C. Mowery and N. Rosenberg (1985), Commercial Aircraft: Cooperation and Competition between the U.S. and Japan. *California Management Review* **27**, 70–92.
[33] D.C. Mowery and N. Rosenberg (1989a), New Developments in U.S. Technology Policy: Implications for Competitiveness and Trade Policy, *California Management Review* **32**, 107–124.
[34] D.C. Mowery and N. Rosenberg (1989b), *Technology and the Pursuit of Economic Growth*, Cambridge University Press, New York.
[35] L.K. Mytelka and M. Delapierre (1987), The Alliance Strategies of European Firms in the Information Technology Industry and the Role of ESPRIT, *Journal of Common Market Studies* **26**, 231–253.
[36] Office of Technology Assessment (1988), *Commercializing High-Temperature Superconductivity*, U.S. Government Printing Office, Washington, D.C.
[37] D. Okimoto (1989), *Between MITI and the Market: Japanese Industrial Policy for High Technology*, Stanford University Press, Stanford.
[38] A. Olechowski (1987), Nontariff Barriers to Trade, in J.M. Finger and A. Olechowski (eds), *The Uruguay Round: A Handbook for Negotiators*, World Bank, Washington, D.C.
[39] S. Phillips (1989), When U.S. Joint Ventures with Japan Go Sour, *Business Week*, 24 July, 30–31.
[40] M.E. Porter and M.B. Fuller (1986), Coalitions and Global Strategy, in M. E. Porter (ed.), *Competition in Global Industries*, Harvard Business School Press, Boston.
[41] C. Sanger (1984), Computer Consortium Lags, *New York Times*, 5 September, D1.
[42] J. Sigurdson (1986), *Industry and State Partnership in Japan—The Very Large Scale Integrated Circuits (VLSI) Project*, Sweden, Research Policy Institute, University of Lund, Lund.
[43] W.E. Steinmueller (1988), Integrated Circuits, in D.C. Mowery, *International Collaborative Ventures in U.S. Manufacturing*, Ballinger Publishers, Cambridge.
[44] J.S. Stuckey (1983), *Vertical Integration and Joint Ventures in the Aluminum Industry*, Harvard University Press, Cambridge, Mass.
[45] G.M.P. Swann (1987), Industry Standard Microprocessors and the Strategy of Second-Source Production, in H.L. Gabel (ed.), *Product Standardization and Competitive Strategy*, North-Holland, Amsterdam.
[46] D.J. Teece (1982), Towards an Economic Theory of the Multiproduct Firm. *Journal of Economic Behavior and Organization* **3**, 39–63.
[47] D.J. Teece (1986), Profiting from Technological Innovation: Implications for Integration, Collaboration, Licensing, and Public Policy. *Research Policy* **15**, 285–305.
[48] L. Tyson (1988), Making Policy for National Competitiveness in a Changing World, in A. Furino (ed.), *Cooperation and Competitiveness in the Global Economy*, Ballinger Publishers, Cambridge.
[49] R.S. Vernon (1979), The Product Cycle Hypothesis in a New International Environment, *Oxford Bulletin of Economics and Statistics* **41**, 255–67.
[50] E. von Hippel (1976), The Dominant Role of Users in the Scientific Instrument Innovation Process, *Research Policy* **5**, 212–239.
[51] O.E. Williamson (1979), Transaction-Cost Economics: The Governance of Contractual Relations, *Journal of Law and Economics* **22**, 233–62.

Chapter 11

Summary and Implications

OVE GRANSTRAND, LARS HÅKANSON AND SÖREN SJÖLANDER

11.1 SUMMARY AND SOME GENERAL TRENDS

The preceding chapters have all dealt with the same specific phenomenon, namely internationalization of R&D and technology, but with different foci and with different empirical data bases. Table 11.1 gives an overview of the different foci, empirical data bases and methods as well as the main findings from the different chapters. However, internationalization of R&D and technology is a phenomenon with many dimensions but with—after all— few available observations so far. This leaves ample room for uncertainty and speculation, and—indeed—anxieties. Although the findings presented here about internationalization of R&D and technology are not clearly conflicting, it is yet premature to make a coherent synthesis and to offer overall conclusions. Instead a simple enumeration of some relevant trends and facts indicated in the preceding chapters will be given below. These trends may also serve as a starting point for formulating some implications for management and policy-making. It is notable that no obvious trends are indicated in the findings regarding the effects of internationalization of R&D and technology. This is a reflection of the current state of knowledge in which we know a great deal of the phenomenon itself and some of its determinants but little in the form of systematic, quantitative evidence of its effects. (Cf. the attempt to relate the degree of internationalization of R&D to direct measures of economic performance in Cantwell's study.) What we know about effects still rests on qualitative evidence and reasoning.

Technology Management and International Business: Internationalization of R&D and Technology.
Edited by O. Granstrand, L. Håkanson and S. Sjölander. © 1992 John Wiley & Sons Ltd

Table 11.1 Overview of the chapters in the book

Chapter	Author(s)	Main focus	Empirical data/method	Main findings
2	Dunning	Overview of causes to, nature of and effects from the generation, organization and diffusion of innovatory capacity through MNCs.	Various official survey data bases (NSF, UNESCO, IMD etc.) plus patent data as in Cantwell plus earlier studies by Dunning plus two case studies of UK motor vehicle and UK pharmaceutical industries.	The effects of MNCs on innovatory capacity of home and host countries cannot be assessed in general or simple terms but depend on a.o. MNC strategies, responses of indigenous firms, institutional structure, and government policies. Case studies give a mixed verdict. Received theory and debate offer two opposing views. Principles of comparative advantage and international division of labor apply to international allocation of innovatory capacity. Increasing internationalization of R&D and technology is likely to continue but with increased pluralism of organizational forms employed.
3	Patel and Pavitt	The relations between large firms' production of technology by sector and country, especially their home country.	US patent data for 1969–86 for 686 of the world's largest industrial firms as they were constituted in 1984, supplemented by firm data. Analysis for 11 countries and 33 technical sectors. Correlation analysis.	Production of technology remains highly domesticized in the sense that foreign R&D does not play a major role in most countries and in most firms and the technological performance of large firms is closely correlated with the technological performance of their home country.
4	Cantwell	Extent and pattern of internationalization of technological activity and its implications for the technological competitiveness of firms and countries.	Same data base as Patel and Pavitt but for 727 of the world's largest industrial firms. Analysis by industry and country. Special focus on UK.	MNCs in UK and in smaller European countries have been considerably more involved in internationalizing research than US and Japanese MNCs, although with mixed implications for competitiveness. With some exceptions the research of large firms has been especially attracted to main centres of innovation for their primary sector of activity. Increasing national specialization is possible.
5	Håkanson	Locational determinants of different types of foreign R&D units (in contrast to company characteristics as determinants).	Approx. 170 foreign R&D units belonging to the 20 largest industrial firms in Sweden. Questionnaire survey. Factor analysis. Linear regressions.	Types of foreign R&D units by dominant motive were: Political, Production support, Market proximity, Monitor research, Multimotive units. Each type had a specific establishment process and set of determinants.

Table 11.1 (cont.)

Chapter	Author(s)	Main focus	Empirical data/method	Main findings
6	Casson, Pearce and Singh	Recent trends in R&D from parent and subsidiary point of view.	Questionnaire survey supplemented with 27 interviews covering one lab in each of 27 firms with lab locations in UK (14), US (4), Europe (6) and elsewhere (3).	Among trends indicated were: Convergence of R&D practice; More diversified R&D; Increased networking among R&D labs.
7	Pearce and Singh	Internationalization of R&D by the world's leading enterprises.	Questionnaire survey of all R&D units of Fortune's 500 largest industrial firms in the world plus 60 additional firms. Responses mainly perceptions on a 3-point scale. Response rate 29%. Analysis by industry and country.	Increased importance of foreign R&D and global approaches to innovation. Three types of foreign R&D units were distinguished: (a) Support laboratory (b) Locally integrated laboratory (c) Internationally interdependent laboratory.
8	De Meyer	Determinants behind internationalization of R&D.	14 case studies of large European (9), Japanese (2) and North American (3) firms. Interviews with research managers and their subordinates.	Technical learning a fundamental determinant behind internationalization of R&D. Increased occurrence of networking.
9	Granstrand and Sjölander	Relations between internationalization and diversification of technology in large firms.	Interviews and questionnaire surveys of 14 Japanese, 16 US and 12 Swedish large corporations, plus a product case study (mobile telephony). Correlation analysis.	Increasing importance during the 1980s of internationalization strategies at firm level and various strategies for external technology acquisition, including internationalization of R&D. Generic technological progress and technology diversification at product and firm level are main driving forces.
10	Mowery	International collaborative ventures for technology commercialization involving US firms.	Draws on earlier studies of multinational joint ventures in commercial aircraft and in US manufacturing, including some case studies.	Internationalization of technology-related collaborations a relatively new and increasingly important phenomenon, spurred by intensified international competition and escalating R&D costs. International collaboration may be a less costly and speedier way of commercializing new technologies and there is no evidence that such collaborations will do great damage to US industry.

Only a partial selection of trends has been made here, without any attempt to be systematic or exhaustive. The trends are naturally interrelated and mostly apply to industrialized countries and to periods spanning at least a decade. Qualifications to the statements are often important, e.g. regarding country, industry and short period variations, but these are left out here. The selected trends have, to the extent possible, been sorted into groups corresponding to Figure 1.1. However, many of the trends or factors underneath the trends interact causally as both determinants and effects. In cases where specific studies identify the trend the name(s) of the author(s) are given within parenthesis.

A. Background (Contextual) Trends

A1a Increased international competition and increased importance of product performance and quality based competition.
A1b Increased differentiation of products and homogenizing of markets internationally.
A2a Increased technology levelling among countries and companies, i.e. technologies are becoming increasingly multi-firm and multi-country led.
A2b Innovations are increasingly having multiple geographical and organizational sources of technology, with increasingly differentiated and innovation-specific patterns of diffusion.
A3a (Mowery, Pearce and Singh) Increasing pressure to shorten international market penetration times for new products and increasing simultaneity in their introduction on various national markets.
A3b (Granstrand and Sjölander) Increasing pressure to shorten R&D times and decreasing market life times for new products.
A4a Escalating R&D spendings (and in some cases also R&D intensity) at product, firm and country levels in industrialized countries.
A4b (Dunning) The technological capacity of developing countries has increased since 1970.
A5   Industrial R&D is becoming increasingly science based.
A6   (Granstrand and Sjölander) Increasing technological protectionism and probably also science protectionism.

B. Trends Regarding Internationalization of R&D

B1   (Granstrand and Sjölander, Pearce and Singh) Internationalization of R&D is with a few exceptions a fairly recent phenomenon. (Only a small part of R&D labs deliberately set up abroad are more than a couple of decades old).
B2a (Pearce and Singh, Cantwell, Dunning, De Meyer a.o.) Increasing internationalization of R&D, although on an average still from a low country- and industry-specific level. (Note that Cantwell's and Patel and Pavitt's patent data base with just one year (1984) of consolida-

Summary and Implications _____ 237

tion of firms' ownership of patents does not accurately pick up this trend because acquisitions cannot be accounted for. Probably the trends towards internationalization of R&D and technology that these studies still show are understated, since acquisitions appear to be a significant factor behind internationalization of R&D, as shown for Sweden by Håkanson among others.)

B2b (Dunning, Granstrand and Sjölander, Pearce and Singh) Internationalization of R&D of increasing importance among Japanese MNCs.

B2c (Dunning) Increasing research intensity of foreign-based production among US and European MNCs.

B3 (Dunning, Granstrand and Sjölander, Mowery) Increasing external and cross-border acquisition (sourcing) of technology among MNCs through various means (acquisitions, joint ventures, licensing, contract R&D, scanning).

B4a (Cantwell, Dunning) Increasing concentration of R&D input resources and R&D outputs (inventions, patents, innovations) to certain (sometimes industry-specific) regions and to certain companies (mainly large MNCs).

B4b (Dunning) Some increase in the geographical dispersion of R&D and patents since 1970. Some convergence since 1970 in R&D and patenting shares among the five leading industrialized countries—US, France, West Germany, UK, Japan. Large sector differences, however.

B4c (Dunning) Many leading MNCs establish R&D units in all three regions North America, West Europe and Japan, while R&D units tend to agglomerate within these regions.

B5 The links between university research and industrial R&D are increasingly strengthened in many industries and also the links between universities and firms of different nationalities.

B6 (Casson et al.) There is a convergence of R&D practice in the sense that many Western firms are moving towards less R and more D, while leading Japanese firms are beginning to put greater emphasis on fundamental research and less on applied.

B7a (Granstrand and Sjölander, De Meyer, Pearce and Singh, a.o.) Evolution of MNCs from 'hub structures' into network- or matrix-oriented organizational structures for international coordination and rationalization of assets, through localized but mobile global responsibilities, among other things.

B7b (Pearce and Singh) Management of R&D and technology is becoming increasingly global in outlook.

B8 Ordered or sequential evolutionary models of international expansion are becoming increasingly insufficient, although not entirely invalid, for describing the strategic behavior of advanced MNCs (e.g. models of gradual expansion to geographically or culturally neighbouring countries, models of sequential internationalization of in turn marketing, production, and R&D functions, or models of centralized generation of technology in an MNC followed by transfer to subsidiaries and adaptation).

C   Trends Related to Determinants

C1   (Håkanson, Pearce and Singh) Output market conditions (product demand factors) still dominate as determinants behind increasing internationalization of R&D. Input market conditions (supply factors) regarding R&D personnel and S&T information are becoming increasingly important, at least in some industries and firms (Granstrand and Sjölander).

C2   (Dunning, a.o.) Internationalization of R&D is closely correlated with and influenced by internationalization of production.

C3   (Cantwell, Casson et al., Dunning, Granstrand and Sjölander, Mowery) Increasing technology diversification at product and company levels and increased interrelatedness among technologies and products as well as among technologies and companies. Technology diversification tends to increase the pressure for external technology acquisition, including international technology acquisition. This tendency would be mutually reinforcing a possible tendency towards increasing technological specialization at country level, as indicated by Cantwell.

C4   Increasing penetration of information and automation technologies in R&D and engineering work, as well as in production, among other things lowering the minimal optimal scale of operations and flattening the optimum in general, thereby allowing increased geographic dispersion of R&D and production operations.

C5   (Mowery) National governments are increasingly able to influence R&D and technology transfer decisions of firms through the use of trade and other policies.

C6   (Pavitt and Patel) An MNC's technological performance is closely correlated with the performance of its home country.

As a concluding comment to the trends above one may point out that despite increasing internationalization, there are still no signs of far reaching denationalization, that is loss of national features. MNCs are still undoubtedly uninational, or in a few cases binational, in the sense that there is a dominant nationality among owners, and/or board members and/or corporate management. Thus all but a few MNCs have a nationality in a fundamental sense (implying the existence of a home country and consequently the relevance of the commonly used distinction between home and host country or domestic and foreign conditions). However, internationalization may lead to some kind of denationalization and the emergence of at least some truly international or stateless MNCs (or with ever-changing nationality) but not necessarily so. In the case it does, the process of denationalization lags considerably behind the process of internationalization. As pointed out by Patel and Pavitt, Pearce and others, there is a long way to go towards genuine globalization of industry and technology.

11.2 MANAGERIAL IMPLICATIONS

In spite of the continued importance of home-based R&D, MNCs are locating increasing shares of their R&D abroad. Moreover, foreign R&D activities are no longer limited to mere adaptations of parent company technology, but increasingly include advanced basic and applied research, new product development and the accumulation of core technological skills.

Concurrently, external sourcing of new technologies has become a cornerstone in many companies' technological strategies. However, efficient external sourcing is not possible without strong in-house technical capabilities. When significant external suppliers of new technology are located outside the country of origin, external sourcing strategies may reinforce the tendency towards R&D internationalization.

Moreover, new technologies are not equally available everywhere. Especially in science-based industries, where universities and other public research institutions are prominent, 'pockets-of-innovation' have emerged, often in highly concentrated geographical areas. Access to these sources may sometimes constitute a critical competitive advantage, adding further inducements to the geographical decentralization also of advanced research activities. The overall significance of such 'supply-side' factors is as yet limited, but appears in many industries to be of growing importance.

Traditionally, home country R&D has formed the hub in an international R&D system, with only a few and generally small foreign R&D units in its periphery (Philips and SKF are among early exceptions). Most significant R&D decisions were taken by headquarters or in home country R&D units, while foreign subsidiaries focused on more routine engineering and design tasks with little or no direct interconnection between the foreign R&D units themselves. In order to avoid the costs and difficulties of coordination, it was kept to a minimum. This was a feasible policy as long as foreign R&D volumes were small and mostly devoted to local market adaptations.

Over time, this dominant structural configuration has increasingly evolved into organizational forms, a 'network model', constituted by specialized foreign R&D sharing the task of developing core technological capabilities, sometimes without a clear superior center. In the 'hub-model', foreign R&D was mostly associated with technical support and local market adaptation. The corporate portfolio of technologies was planned and controlled in the home country. Reverse (from subsidiary to parent) and lateral (between subsidiaries) flows of technology were rare. In contrast, the network model is associated with decentralized planning and control, technological specialization and diversity. Many units contribute to the technological competitiveness of the corporation. In a simplified way one may state that while the 'hub-model' was primarily based on the exploitation of economies of scale, the 'network model' obtains its significance from its ability to reap economies of scope.

On the firm level, geographical and organizational decentralization of R&D creates new needs for coordination, communication and control in order to ensure the direction, efficiency and effectiveness of technological progress. Systems, procedures and organizational structures must be designed to overcome cultural, organizational and geographical barriers in order to ensure cooperation and a smooth flow of information between units. As indicated in several chapters in this book, there is a great need for enhancing our current understanding of these issues. A more systematic empirical study of current organizational practice is needed before we can reach firmly grounded normative conclusions.

Nevertheless, the uncertainties inherent in technological development inhibit the effectiveness of formal instruments for planning, coordination, control and information exchange. Successful innovation often requires new combinations of skills and expertise that are difficult or impossible to foresee. Here, informal personal networks play a fundamental role. For this reason—and because they play a vital role in facilitating information exchange built on mutual trust and respect—measures such as corporate-wide conferences and rotation of personnel are increasingly used to promote the development of informal networks spanning countries, cultures, functions and scientific/technical disciplines.

The evolution of multinational companies towards increasingly intricate network structures entails new and central managerial challenges. These concern not only the planning, coordination and 'logistics' of large and complex flows of technology, finance, people and materials but, more fundamentally, the nature of the process by which strategy is formulated and implemented. The strategic management of technology in an integrated network structure, characterized by wide geographical and organizational dispersion of technical capabilities and resources, assumes special significance. The setting of a long-term strategic direction for the evolution of core corporate technologies is hardly ever amenable to formal, traditional long-term planning techniques, nor does the pursuit of new technical capabilities easily lend itself to formal instruments of control and supervision. To a significant extent, integration of strategic efforts must instead rely on shared values and visions as expressed in strong corporate cultures, a widely shared awareness of central goals and strategies, etc. Hedlund [6, p. 24], uses the metaphor of the 'holographic corporation':

Summary and Implications

The distribution of information in each part of a hologram is possible because of laser technology. One could say the corporate ethos is the analogue of the laser light. By sharing certain conceptions about the firm, and certain ways of acting in relation to other members of the firm, it becomes possible to rapidly share information, interpret the meaning of events in and outside the organization in similar ways, and see opportunities for local action in the interest of the global good. The laser beam effect of corporate culture is the unifying element in an heterarchical organization. It is crucial to support the formation of such a culture, since the risks of anarchy are otherwise very great.

Shared values and a common vision do not by themselves solve the problem of coordination and planning in an R&D network, but they help to give some coherence in R&D activities. In parallel with increased geographical decentralization, increased centralization of the control of technology development is needed in many firms to ensure that important technological developments are taken care of and that overlapping R&D activities are avoided. Moreover, the R&D and technology planning processes of multinational R&D activities must be linked to the corporate business planning. For instance, at the corporate level the strategy of product diversification to spread R&D costs and business risks as an alternative or complement to market diversification is intimately linked to the management of corporate technological assets.

Internationally decentralized and dispersed R&D means that the need for centralization of the control over the corporate technology portfolio increases. International coordination means not only having corporate-wide information on the technology portfolio and R&D activities and managing different professional subcultures but also bridging different corporate subcultures as well as country cultures. International coordination also means that people have to meet to exchange ideas and information. Corporate technology management has to establish ways and means also for informal coordination and international training.

Escalating R&D spendings have been discussed in previous chapters. Escalating R&D spendings increase the pressure on technology managers to be more cost conscious and to find ways and means to balance the R&D cost increase by means of out-sourcing, partnering, reorganization of R&D and rethinking the way product development work is carried out to reduce lead times *and* costs. Among factors behind increased R&D spendings are increased technology diversity on product, firm and industry levels. One of the keys to control the cost effects of technology diversification is efficient integration of technology. A dominating prerequisite for this is a multi-skilled engineering workforce and a primary means is job rotation. It is our impression that too many Western companies practise too little job rotation and that they have a great deal to learn from best practice companies, many of which are Japanese.

Industrial R&D is becoming increasingly science-based, especially in areas such as biotechnology, information technology and new materials. In the future the links to universities and other science centers are going to play a fundamentally more important role. In order to be able to develop such links and to participate in national as well as international science networks, companies have to have receiver capacity. But more fundamentally, in order to be able to enter scientific networks, companies have to offer something besides money. In this respect corporate research plays a very important role, a role sometimes neglected by firms. Good university scientists can seldom be bought with money but links have to be built on mutual professional as well as personal respect and on the formulation of challenging research questions.

11.3 POLICY IMPLICATIONS

The preceding chapters have directly or indirectly raised a number of policy issues and, as is always the case, there are more questions than the world can answer. Although policy analysis is not the prime purpose of this book, a few policy implications will be pointed at here.

A general implication is that there is little doubt that internationalization of R&D and technology is an area of legitimate policy concern, especially for national governments but also for supra-national policy-making bodies. The general policy concern about internationalization and international competitiveness of national economies and firms is well known. The decisive importance of R&D and technology for economic performance gives additional reason to be concerned specifically about internationalization of R&D and technology. The cases of Korea and especially Japan also show what activistic and coordinated national policy-making can do for reaping the national economic benefits of technology, as Dunning has pointed out in this volume. (See also [4].) The current notion of 'national systems of innovation' as developed by Freeman, Lundvall, and Nelson (see[3]) is in line with the belief that national institutional structures, policy environments, industrial organizations, traditions, etc. impact economic performance. This is in line with the results of e.g. Patel and Pavitt in this volume as well as with Porter [7].

The question is what role the two components—foreign-located but domestically owned R&D, and domestically located but foreign-owned R&D—can play for the performance of a national system of innovation and for a national economy at large. A sequel question is how this role can be affected by policies, both of the 'Ordnungspolitik' kind (affecting institutional structure, legislation and judiciary matters for shaping and maintaining the economic rules of the game) and the 'Prozesspolitik' kind

Summary and Implications 243

(intervening in the game). The questions are complicated by the intangible, risky, and long-term nature of R&D and technology, by the mobility of R&D and technology resources across geographical and organizational borders, and by the frequent influence of small events, early moves, increasing returns on technological activities, and lock-in effects. Arthur [1] elaborates the latter point, which has relevance also for locational clustering into hi-tech areas as well as for technological specializations of nations and regions.

These features of R&D and technology are in themselves sufficient reason for policy concern regardless of the question whether internationalization of R&D and technology has proceeded very far or whether there is a trend towards increasing internationalization or not. However, the trends indicated in the preceding chapters and summarized above are fairly clear and point, among other things, at increasing concentration of ownership and geographical dispersion of innovative capabilities, although with national and industry variations. If the available evidence is extrapolated, internationalization of R&D will indeed have become significant by various measures within a period spanning, say, two decades ahead. A period of 20 years corresponds to the lifetime of a patent and is of the same magnitude as the time needed to reach break-even for many investments in new technologies. Thus, a 20-year period is not a too long-term horizon for technology planning and policy purposes.

A second policy implication is that *general* policy measures with the objective to promote, inhibit or otherwise influence foreign direct investments in R&D and technology cannot be tailored with sufficient precision, at least not on the present grounds of knowledge about internationalization of R&D and technology. For one thing, corporate investment decisions affecting R&D location are usually made considering other factors than merely R&D, e.g. marketing and production, or they are made in a larger context, e.g. of an acquisition, as Håkanson has described in this volume.

Moreover, as Dunning in particular has demonstrated, there is considerable variation in the situational contingencies regarding different countries, technologies, sectors and firms, and their strategies and responses, and the different motives for and types of foreign direct R&D investments and their subsequent evolutionary patterns. Nothing but a mixed verdict can then be given regarding the home and host country effects in general of the interaction between foreign R&D and a national system of innovation. However, as Dunning and others also have shown, theoretically as well as empirically, there are specific situations where the net effects could be expected to become positive for both home and host countries. An example is when foreign R&D interacts with a home and a host system of innovation and a home and a host economy at large that is sufficiently well developed to be capable of reaping positive allocative effects, positive externalities and other economic rents. To the extent that such a local capability is absent, say in some industry at the host country end, but could be built up with expected long-term economic viability, there is in principle a case for *temporary* infant industry protection (or 'infant innovation system' protection). The effective functioning of such policies for building up parts of a national system of innovation would be jeopardized if foreign direct investments in the corresponding industry and R&D areas have already reached a significant level or if such policies lead to asymmetries among nations in the access to product and factor markets which would result in reciprocity claims and retaliatory action.

There are also situations when foreign R&D investment would result in negative net effects for home and/or host country. An example of negative host country effects would be when an MNC acquires technology and R&D resources in a foreign country for less than their local opportunity cost and uses these resources to outcompete the local industry with no positive restructuring effects on the local economy. This could for instance happen if the host country supply of qualified scientists and engineers is scarce and local competitors are small but growing. This does not necessarily have to take place in a country with a generally weak S&T infrastructure and a weak national system of innovation. The acquisition of new, small technology-based US firms by large Japanese MNCs could be a case in point, especially since similar opportunities do not, for various reasons, exist to the same extent for US MNCs in Japan (see [4]). Although lack of symmetric access or lack of reciprocity is a legitimate reason for policy concern, it is on the other hand to be expected that differences in national S&T infrastructures and innovation systems naturally give rise to some asymmetries which cannot in all cases justify policy action. Moreover, as shown by Granstrand and Sjölander [4], large MNCs could acquire small, technology-based firms in ways that significantly stimulate the subsequent growth and innovativeness of the small firm within the organization of the larger MNC. This is a further reason why caution should be exercised when formulating general policy recommendations in this context. At the same time this points to the general need to develop methodologies for policy analysis and policy evaluation that take situational variations sufficiently into account.

Thus, in summary of this point, there are situations in which foreign R&D investments create positive sum games between MNCs, home and host countries, and others where zero sum games, and conceivably even negative sum games, are created and the situational specificity precludes effective general policies. This is even more so in countries with industries where internationalization has proceeded far.

A third implication for policy-making concerns the limited relevance of supply-side policy measures regarding internationalization of R&D and technology. A common experience with many such measures for stimulating R&D and innovativeness in general is that their effectiveness in industri-

Summary and Implications

alized countries with a fair number of large corporations is limited, unless they are combined with demand-side measures (e.g. for technology procurement). For example, R&D tax credits and other R&D subsidies have by themselves very limited effects, partly because they are launched on a limited scale and do not tie strongly into corporate decision-making regarding R&D. This does not mean that supply side measures could not be tailored to work at all. A double tax deduction scheme for R&D costs, as was operating in Singapore in the early 1980s, represents quite another incentive level than a R&D tax credit system like the one introduced in the USA in 1981 and the R&D tax deduction system abandoned in Sweden in 1983. Similarly, tax deductions for qualified R&D personnel salaries and repatriation schemes for R&D personnel could be made quite effective in stimulating mobility and recruitment of special R&D personnel.

In connection with internationalization of R&D and technology, Pearce, Håkanson and others have shown, moreover, the limited relevance of supply-side factors for locating foreign R&D. On the other hand, the cases where supply-side factors matter seem to become more frequent. At the same time such cases are associated with a type of R&D activity that from the host country point of view should raise the most policy concern (that is IILs in Pearce's terminology and Type IV R&D in Dunning's terminology). In order for such R&D activities to give positive host country effects, their suitable interaction—directly or indirectly through related production or marketing—with local industry and local S&T infrastructure must be ensured. As has been pointed out above, and especially by Dunning, a prerequisite for mutual beneficial interaction to take place is that the national S&T infrastructure and the whole national system of innovation have a sufficient degree of development and strength. To accomplish this is indeed an important task for national policy-making, which involves building up effective institutions for S&T education at all levels, to build up local S&T capabilities to scan, acquire, absorb, refine and exploit foreign R&D and technology, to sustain frontier research capabilities in some areas—not least to provide a pattern and a 'ticket of admission'—to provide effective mechanisms for domestic technology transfer and to provide an environment conducive to technology-based innovation and entrepreneurship. S&T policy-making made with these objectives in turn involves policy-making in many areas (labor, finance, education, trade, industry, intellectual property, etc.). This points to the need for coordinated, coherent policy-making regarding technology, which should come as no surprise to those familiar with the pervasive nature of technology and the imperfections of markets in dealing with it.

A fourth implication is somewhat unsurprising but nevertheless important, and that is the need for further policy research regarding internationalization of R&D and technology. As has been pointed out in several places in the preceding chapters, research in this area is only recent and has not attracted much interest among economists, scholars of MNCs and other researchers. Much work is called for in order to develop relevant measures and methodologies for policy analysis as well as to build up databases to allow cross-country comparative studies. How research may suddenly be fruitfully spurred by new rich databases made available is demonstrated by the databases built up at SPRU and the University of Reading and drawn upon by several of the authors in this volume. At the same time more data of these and other kinds are needed to check and expand research results in an emerging area of research with still much paucity of data and case observations.

Finally, an increasing nationalistic policy concern about internationalization of R&D and technology is far from unproblematic. First of all such concern may result in ill-conceived xenophobia and inconsistent fears for both outward and inward R&D investments and migration of R&D personnel. Overly nationalistic policies may also create conflicts with the traditional international orientation and openness of large parts of the S&T community. But more importantly, nationalistic policies could simply result in undesirable protectionism. To the extent that nations engage in technology-based international competition, which in fact they do and there are reasons for their doing so, claims to national short-run appropriation of sufficient benefits of nationally financed investments in S&T, e.g. tax financed public R&D and education at publicly owned universities, become legitimate. It then becomes difficult to distinguish between 'reasonable' national appropriation and national protectionism, which may be detrimental to the nation itself in the long run and/or may be directly or indirectly (through retaliation) detrimental to other nations—or protectionistic trade blocks or continents for that matter. It also becomes of the utmost policy concern at a supra-national level to mitigate such detrimental effects of protectionistic national or regional policies rather than to rely on dispute settlements through scattered bilateral agreements, which tend to be too narrow in scope. Ultimately, internationalization of R&D and technology and the functioning of international factor markets are of international policy concern, just as international trade and the functioning of international product markets are. Ensuring effective international competition on the latter type of markets is important and will probably in the not too distant future call for some kind of international anti-trust policies in the light of some ongoing concentration and cooperation tendencies. Similarly, international 'Ordnungspolitik' to ensure some forms of international competition or contestability on national factor markets for R&D and technology may be called for to correct severe and persistent national asymmetries or S&T protectionism, of which there are several current examples.

11.4 A SPECULATIVE OUTLOOK ON THE FUTURE

In the light of the pervasiveness of technological change and the importance of internationalization, it seems appropriate to conclude this book with some speculations about the future.

First of all, hopes were raised in the early 1990s for a decreased role of the military industry and military R&D in the world. Military R&D has at times roughly equalled half of the world's R&D and has moreover been highly domesticated, as has the military industry. (Except for military aid the international arms trade consists to a considerable degree of exports from nations that want to rely on and sustain a sizable national defense industry.) Thus a large component of national systems of innovation and national economies may become smaller and perhaps less subjected to national concerns. Relatively seen, internationalization of R&D and technology would thereby increase as would the share of privately funded R&D, but the real question is what will happen to military R&D and the possibly released R&D resources. Civilian technology has taken the lead over military technology in many instances, and this will probably become even more frequent in the future, as will instances of 'reverse spill-overs', and there are still possibilities for dual civilian/military uses of new technologies, although in some areas civilian and military technologies are claimed to diverge. Moreover, technology diversification, rising R&D costs and diminishing possibilities of nations becoming technologically self-sufficient apply to military products as well. Thus, it is conceivable that military R&D and technology will also be increasingly internationalized in the future. The implications of this for military policy-making and international security are left to the reader to speculate about.

A second component in national systems of innovation that has not been internationalized to any significant degree is the university system. It is true that many universities, especially in the US and the UK, export a sizable amount of teaching through their foreign students. Many universities have foreign R&D contracts, although mostly small, and some universities have overseas campuses. However, for practical purposes one cannot yet talk about the existence of multinational universities, that is with a new acronym, MNUs, but one can speculate on the emergence of such universities (or families of allied universities for that matter). The first MNUs will most likely be US universities with international prestige. As an institutional innovation the university is much older than the industrial firm. Moreover, universities were in most cases not originally created solely on the basis of regional or national economic motives, although regional economic motives played a role in the creation of US state universities. Thus, they could conceivably be replaced by more modern and economically efficient forms of organizing higher S&T education and R&D for sustaining national competitiveness, as such objectives come more to the forefront on national scenes. On the other hand, it is possible that a university with a good reputation and standing could develop a viable business idea which would make it natural for it to set up foreign subsidiaries, to draw on national S&T infrastructures abroad and to cater to foreign markets for teaching and research, either by itself or through joint ventures. A degree from an MIT subsidiary in Singapore would still be an MIT degree but without the expenses for the Singaporean student incurred by going to MIT in the US.

The upper league of US universities is claimed to be strong and efficient in international comparison and important for the international competitiveness of US industry (see [2]). If so, and disregarding US nationalism for the moment, they could in principle expand their business abroad, e.g. in Asian countries from which they source many of their students and some of their faculty. Analogously to the traditional expansion pattern of MNCs they could start with subsidiaries for marketing and production—that is teaching—and then eventually move into foreign research as well, in the first stage, however, only to support local teaching and PhD-production.

Thus a case could be made for the emergence of MNUs, which would further the internationalization of R&D and technology and the international interlocking of national systems of innovation.

Finally, one can speculate about the future role of nation states and the prospects for 'denationalization' of various S&T institutions. Nation states and their underlying ideas are just a few centuries old (originating in modern form in connection with the American and French revolutions in the 1780s), younger than many S&T institutions and ideas about S&T while older than the industrial revolution and its main institutional innovation—the industrial firm or, more generally, the business firm. The latter has during the last century or so displayed a remarkable versatility and strength in its evolution, producing varieties such as large MNCs, MPCs and MTCs with important features, such as the industrial R&D laboratory and the internal corporate bank. As an institution, the business firm shows no sign of evolutionary decline, quite on the contrary. It is thus conceivable that the business firm, equipped as it is with economic power and technological and managerial competence and trimmed by competition, will further evolve and take on wider responsibilities, including some responsibilities that have traditionally been assigned to national governments and public bodies. In relation to large corporations, it is not unlikely that individual nation states will have fewer roles to play in the future and become less and less important, not least regarding funding and performance of R&D, as internationalization and concentration proceed. It is further conceivable that the same competition that gives the business firm its strength allows it to grow to the point where competition is dulled by too much concentration of industry and R&D into the hands of too few firms for effective competition to be sustained. However, such a situation can hardly be influenced by a single nation acting in isolation. Justified as they may be as regional economic actors, nation states also have to link up with each other, internationalize and take on a global outlook in order to avoid becoming marginalized parochialists. Technology flies no flag.

REFERENCES

[1] W.B. Arthur (1988), Competing technologies: an overview, in Dosi et al. (eds), *Technical Change and Economic Theory*, Pinter, London.
[2] M. Dertouzos, R. Lester and R. Solow (1990), *Made in America. Regaining the Productive Edge*, Harper Perennial, New York.
[3] G. Dosi, C. Freeman, R. Nelson, G. Silverberg and L. Soete (eds) (1988) *Technical Change and Economic Theory*, Pinter, London.
[4] C. Freeman (1987), *Technology Policy and Economic Performance. Lessons from Japan.* Pinter, London.
[5] O. Granstrand and S. Sjölander (1990), The Acquisition of Technology and Small Firms by Large Firms, *Journal of Economic Behavior and Organization*, **13**, 367–386.
[6] G. Hedlund (1986), The Hypermodern MNC—A Heterarchy? *Human Resource Management* **25**, 9–35.
[7] M.E. Porter (1990), *The Competitive Advantage of Nations*, Macmillan, London.

Index

acquisition of technology-based firms 189
acquisition through external cooperative R&D 189
acquisitions 9, 101
adaptation and improvements 29
Airbus 227
Airbus Industrie 210
Alfa-Laval 187, 188
Alfred Nobel 189
Allen, T.J. 176
Allied 56
ambassador role 175
Anglo–French Concorde partnership 228
appropriation 246
Arthur, W.B. 4
Astra 187
AT&T 56, 194, 210
Audretsch, D.B. 78

Bartlett, C.A. 4, 140
basic materials or product research 30
Bayer 55
Behrman, J.N. 170
Bell Labs 193
Boeing 216, 227
Boliden 187

CAD/CAM 177
Canon 55
Cantwell, J. 54
Caves, R.E. 11
cellular radio 193
collaboration 210
collaborative ventures 209
  costs and benefits of 225
  management of/dynamic character of 224

structure of 223
types of 211
communication 166, 170, 176
concentration 239
  of R&D input 237
Concorde 227
control 171, 240
convergence of R&D 237
cooperation 172
coordination 166, 203, 240
  and control 7
core corporate technologies 240
corporate development strategies 191
cross-boundary contacts 172

De Bodinat, H. 170
De Geus, A.P. 173
decentralized R&D 241
differentiation of products 236
direct foreign investment 215
diversification 102, 190
  of technology 182
documentation 176
Dow 55
Doz, Y. 227
Du Pont 56
dynamic character 224

economies of scale and scope 203
EF Johnson 194
electronic communication 176
environmental trends 184
Ergas, E. 44
Ericsson 192
  Radio Systems 192
  Telecommunications 192
escalating R&D spending 236
exploitation of technology
  firm-specific assets 214

external acquisition of technology 237
external networks 175
external sourcing 239
  of technology 237

Fairchild 226
  Semiconductor 222
Fiat 218
Fischer, W.A. 170
Fokker 226
foreign innovating activities 30
foreign production ratio 82

General Electric 55, 216, 227
General Motors 216, 222
generic technological progress 199
geographical dispersion of R&D 237
geographical distances 176
Ghoshal, S. 4, 140
government regulations 9
Griliches, Z. 78
GSM 195, 198

Hagedoorn, J. 213
Hall, B.H. 78
Hamel, G. 227
Harris 194
Hedlund, G. 240
Hewitt, G. 11
Hirschey, R.C. 11
Hitachi 55
Hoechst 56
hub-model 239

IBM 55, 210
Iggesund 187
infant industry protection 244
informal personal contacts 176
innovation capacity
  by country 20
  by firm 25
innovative activity 83
innovatory capacity—government policy 43
input market conditions 238
integration 176
internal network 174
International Aero Engines 218
international collaborative venture, definition of 211
international competition 199, 236

international division of labour 76
international location 83
internationalisation 65, 75, 182
  of R&D, a model 101
  of R&D, general trends of 233, 236, 238
  of production 238
  of technological activity 77, 89
intra-company networks 175
Ishikawajima-Harima Heavy Industries 218
ITT 194

Jacobeus, C. 192
Japan Commercial Transport Development Corporation 216, 227
joint ventures 212
  causes of 218
JT10D jet engine project 226

Kawasaki Heavy Industries 218
Kema Nobel 187
Kodama, F. 197

lateral flows of technology 239
learning 169
  definition of organisational learning 169
licensing 211, 213, 214
listening posts 72
local network 174
Locally Integrated Laboratory 141

Magnetic 194
managerial implications 239
managerial issues 184
managerial perceptions 186
market diversification 182, 201
market penetration 219
market size 106
Matsushita 56
MCC 217
McDonnell Douglas 218, 226
MDF100 commercial aircraft project 226
Meyers, P.W. 173
MIT 210, 248
Mitsubishi Heavy Industries 218
MNCs, network character of 3
MNU 247
Mobil 55

Index

motor vehicle industry 39
Motoren-Turbinen Union 218
Motorola 175, 194, 227
multi-technology corporations 182
multinational coordination 188
multinational universities 247
multinationality, degree of 83

national competitiveness 89
national specialisation 84
National Steel 216, 222
national systems of innovation 71
nationalistic policies 246
NEC 175, 194
network 170, 174
  model 239
networking 173
new entrants 224
Nippon Kokkan 216, 222
Nissan 55
NMT 450 195, 198
NMT 900 195
nontariff barriers 222
nontariff restrictions 222

organisation structure of R&D 237
Oskarsson, C. 192
output market conditions 238
Overseas Subsidiary Laboratories 142

Pakes, A. 78
Parent laboratories 142
patenting
  by large firms 63
  by nationality 82
  country performance 64
  firm and country specialisation 69
  foreign firms by industry 80
  foreign R&D units, by technological activity 81
  foreign, UK firms by industry 90
  in the US by firms in UK, Germany and France 91
  large firm performance 64
  nationally-controlled firms 67
  sectoral performance 68
  US data 53
  US sectoral pattern 60
penetration 219
pharmaceutical industry 39
Philips 56, 188, 239

Phillips, A. 227
planning 171, 240
  process 170
pockets-of-innovation 239
policy 242
  implications 242
  making 242
political factors 105
Porter, M.E. 4
Prahalad, C.K. 227
Pratt and Whitney 218, 226
product cycles 219
product diversification 102, 199
productivity 166
protectionism 246

quasi-integration 204

R&D
  adaptive 102
  centralization of 6
  coordination of 145
  costs 219
  current trends in, by headquarters 120
  current trends in, by subsidiaries 121
  decentralisation of 83
  decentralization 8, 9
  expenditures 20
  generic 102
  global dispersion of 143
  globalised 71
  government funding of 131
  government policy 130
  horizontal acquisition of 102
  host country impact 34
  hypotheses for local determinants 106
  importance of large firms 71
  influence 150
  international competitiveness 94
  internationalisation of, organisation and motivation 137
  internationalization of 239
  lead times 236
  location 106
  location, determinants of 112
  location, host country scientific infrastructure 107
  location, need for market adaptation 107

R&D (cont'd)
  location, political factors 108
  location, psychic distance 109
  locational determinants 105
  locational patterns 32
  management practice 125
  market proximity 103
  monitor research 104
  networks 170
  organisation structure 123
  organisation structure by headquarters respondents 126
  organisation structure by subsidiary respondents 126
  political factors 102
  product diversification 102
  science based 236
  scope of corporate 118
  spendings 241
  strategies 209
  subsidiary influence, by industry 150, 151
  types of 29
  units, typology of foreign 100
rationalized research 30
RCA 55
research collaboration 210
Research Triangle Park 195
revealed patenting advantages 24
revealed technological advantage 84
reverse spill-overs 247
Riordan, M.H. 204
Rolls Royce 218, 226

Saab 226
Schakenraad, J. 213
Science Policy Research Unit (SPRU) 78
SF340 commuter aircraft project 226
Shockley, W. 193
Siemens 56
SKF 187, 188, 189, 239
SNECMA 216, 227
sources of technology 236
sourcing 196
  external, international 199
spillovers 221
startup firms 221
stateless corporation 54
strategic alliances 72
subsidiary R&D
  growth determinants 159
  sources of project ideas by host country 157
  sources of project ideas, by industry 156
  units, involvement by parent and sister 146
support laboratories 140
Swedish multinationals 97

technical learning 168, 169
technical standards 220
technical support laboratories 8
technological activities/local differentiation/geographical dispersion 77
  size of 83
technological capacity 236
technological competitiveness 83
technological convergence 219
technological opportunities 204
technological progress 198
technological specialisation 83, 92, 238
  in pharmaceuticals by country 93
technology
  acquisition 187
  diversification 196, 198, 199
  levelling 236
  purchasing through licensing 189
  scanning 190
  transfer 216
Toshiba 55, 227
Toyota 216, 222
trade barriers 221
trade policies 238

United Technologies 56
University of Reading 78
user–supplier collaborations 223

V2500 jet engine 218
ventures 210
Vernon, R. 54
video-conferencing 177
vision 241
Volvo 187

Westinghouse 55
Wilemon, D. 173
Williamson, O. 204